Livestock Development in Kenya's Maasailand

Westview Replica Editions

The concept of Westview Replica Editions is a response to the continuing crisis in academic and informational publishing. Library budgets for books have been severely curtailed. Ever larger portions of general library budgets are being diverted from the purchase of books and used for data banks, computers, micromedia, and other methods of information retrieval. Interlibrary loan structures further reduce the edition sizes required to satisfy the needs of the scholarly community. Economic pressures on the university presses and the few private scholarly publishing companies have severely limited the capacity of the industry to properly serve the academic and research communities. As a result, many manuscripts dealing with important subjects, often representing the highest level of scholarship, are no longer economically viable publishing projects--or, if accepted for publication, are typically subject to lead times ranging from one to three years.

Westview Replica Editions are our practical solution to the problem. We accept a manuscript in camera-ready form, typed according to our specifications, and move it immediately into the production process. As always, the selection criteria include the importance of the subject, the work's contribution to scholarship, and its insight, originality of thought, and excellence of exposition. The responsibility for editing and proofreading lies with the author or sponsoring institution. We prepare chapter headings and display pages, file for copyright, and obtain Library of Congress Cataloging in Publication Data. A detailed manual contains simple instructions for preparing the final typescript, and our editorial staff is always available to answer questions.

The end result is a book printed on acid-free paper and bound in sturdy library-quality soft covers. We manufacture these books ourselves using equipment that does not require a lengthy make-ready process and that allows us to publish first editions of 300 to 600 copies and to reprint even smaller quantities as needed: Thus, we can produce Replica Editions quickly and can keep even very specialized books in print as long as there is a demand for them.

About the Book and Author

Livestock Development in Kenya's Maasailand: Pastoralists' Transition to a Market Economy

Phylo Evangelou

Throughout Subsaharan Africa, traditional pastoral societies are experiencing great change as ecological limitations intensify and national demands for livestock products expand. In light of these pressures, the transition of pastoralists from a principally subsistence way of life to greater involvement in market economies seems inevitable. Focusing on the Maasai society of southern Kenya, Dr. Evangelou examines factors hindering this transition and discusses possibilities for facilitating positive change. The activities of producers representing different tenure systems and ecological settings are investigated, as is the performance of livestock traders and butchers. Dr. Evangelou reviews development policies previously implemented in Maasailand and concludes by recommending measures for increasing livestock production to meet future needs.

Dr. Phylo Evangelou, an agricultural economist with Robert R. Nathan Associates, is project evaluation economist on the Irrigation Management Systems Project in Egypt and co-editor of *Livestock Development in Subsaharan Africa: Constraints, Prospects, Policy* (Westview, 1984).

To Jim Simpson,
in gratitude and friendship

Livestock Development in Kenya's Maasailand

Pastoralists' Transition to a Market Economy

Phylo Evangelou

Routledge
Taylor & Francis Group
LONDON AND NEW YORK

The field research for this book was conducted in association with the International Livestock Centre for Africa (ILCA), which is part of the CGIAR system comprising thirteen international agricultural research centers. It has two key tasks: to apply existing knowledge to improve livestock production in Africa and to undertake research to fill major gaps in that knowledge. The centre is based in Ethiopia and operates throughout Sub-Saharan Africa. It has a multidisciplinary professional staff of about 80 scientists.

First published 1984 by Westview Press

Published 2018 by Routledge
52 Vanderbilt Avenue, New York, NY 10017
2 Park Square, Milton Park, Abingdon, Oxon OX14 4RN

Routledge is an imprint of the Taylor & Francis Group, an informa business

Library of Congress Cataloging in Publication Data

Evangelou, Phylo.
Livestock development in Kenya's Maasailand.

(A Westview replica edition)
1. Masai--Economic conditions. 2. Masai--Social conditions. 3. Animal industry--Kenya. 4. Herders--Kenya. 5. Economic development projects--Kenya. 6. Rural developoment--Kenya. I. Title.
DT433.545.M33E9 1984 338.1'76'00967623 84-15195

ISBN 13: 978-0-367-02006-4 (hbk)

Contents

Kenyan Shilling—U.S. Dollar Exchange Rates, 1972 to 1982

Year	Kenyan Shillings Equivalent to $1.00
1972	7.1429
1973	7.0012
1974	7.1429
1975	7.3432
1976	8.3671
1977	8.2766
1978	7.7294
1979	7.4753
1980	7.4202
1981	9.0475
1982	10.9223

Source: International Monetary Fund (1983).

Abbreviations and Acronyms

AAME	active adult male equivalent
AFC	Agricultural Finance Corporation
AI	artificial insemination
ALDEV	African Land Development program
ALMO	African Livestock Marketing Organization
AU	animal unit
CBPP	contagious bovine pleuro-pneumonia
ECF	East Coast fever
GDP	gross domestic product
ha	hectare
ILCA	International Livestock Centre for Africa
kg	kilogram
KGR	Kajiado Group Ranch
KIR	Kajiado Individual Ranch
KLDP	Kenya Livestock Development Project
km	kilometer
KMC	Kenya Meat Commission
Ksh	Kenya shilling
LE	livestock equivalent
lit	liter
LMD	Livestock Marketing Division
mm	millimeter
mt	metric ton
RMD	Range Management Division
S-C-P	structure-conduct-performance
SEAZ	Small East African Zebu
SSU	standard livestock unit
W&M	White and Meadows [1981]
WNP	Western Narok Producer

Acknowledgments

This work could not have been accomplished without the generous help of members of the Kenya Country Programme of the International Livestock Centre for Africa. Many individuals contributed to the success of the field research, from drivers and office personnel to the Country Team Leader, Solomon Bekure. The selfless assistance given me by Paul Chara and Peter Lembuya Ole Kapiani deserves special recognition.

James R. Simpson of the Food and Resource Economics Department, University of Florida, provided critical guidance at the each of the study. I am also grateful to Hugh Popenoe and Chris Andrew of International Programs, Institute of Food and Agricultural Sciences, University of Florida, for granting me financial assistance, including funds for travel to and from Kenya.

A special note of thanks is due Adele Koehler for the care with which she typed the tables and bibliography, and Kim Feigenbaum, for preparing the excellent set of maps.

Finally, heartfelt appreciation is extended to Kenyans with whom I associated, especially the Maasai producers, for the forbearance with which they accepted my research inquiries and, in so many instances, for offering the hand of true friendship.

Phylo Evangelou

1 Introduction

Two facts are immediately confronted when one considers the current economic situation and development prospects of Subsaharan Africa (South Africa excluded). First, it is one of the poorest regions of the world, with a per capita annual income of only $329 in 1979 [IBRD 1981]. Secondly, Subsaharan Africa's population is the youngest (45 percent under 15 years of age) and fastest growing (3 percent per year) of any region in the world [UN, Department of International Economic and Social Affairs 1981]. Low levels of production and high population growth rates have combined to severely hinder the region's agricultural development. During the 1970s food production actually declined 1.1 percent per capita annually, and an estimated one-third of the people suffer malnutrition due to inadequate food supplies [Lofchie and Commins 1982; Walters 1982].

The livestock industry, in particular, has recorded one of the most unsatisfactory performances of any subsector of the African economies. The continent contains one-fourth of the world's grasslands and one-eighth of its cattle, but beef production per hectare (ha) is only one-fifth and milk production one-tenth of the world averages. Average productivity per animal is equally discouraging, with beef production two-fifths and milk production one-fifth of the world averages [Crotty 1980]. It is not surprising that meat production per capita is decreasing when these low levels of production are considered in conjunction with the continent's rapid population growth rate. By the year 2000, Subsaharan Africa's estimated meat deficit will equal three million metric tons--68 percent of total production (including poultry and pork) in 1980--if present trends continue [FAO 1982a; ILCA 1980a]. Livestock inventories will need to double by the year 2000 simply to maintain current per capita consumption levels, assuming there are no increases in animal productivity [Simpson 1984].

In the midst of such dismaying conditions, the underexploited productive potential of range areas, regions mainly utilized by largely self-sufficient, nonmarket oriented pastoral peoples, is increasingly attracting the attention of national governments [Institute for Development Anthropology 1980; Ruthenberg 1980a]. The nutritional needs of burgeoning populations make the expansion of livestock production in these areas imperative [Ayre-Smith 1976; Unesco/UNEP/FAO 1979], and yet pastoral systems are characterized as "production systems in a waiting room of development," the presumption being that "development must be expected to set in elsewhere" [Jahnke 1982, p. 103]. The challenge lies in disproving this presumption by overcoming the constraints which currently prevent expanded livestock production in pastoral areas. One of the countries faced by this challenge is Kenya, where a principal focus of livestock development efforts is the pastoral region called Maasailand.[1]

THE PROBLEM

Kenya's Livestock Sector

Slowed growth and expanding demand. Kenya is situated on the east coast of Africa and has a land area of about 570,000 sq km, nearly the size of Colorado and New Mexico combined (Fig. 1.1). Its reputation among the nations of Subsaharan Africa as having a relatively healthy economy is by and large well-earned, but problems have mounted in recent years. A decade of commendable advance following the country's independence 20 years ago has been succeeded by a second decade of economic stagnation.

Between 1964 and 1972, Kenya's gross domestic product (GDP) at constant prices grew at a rate of 6.8 percent per year, with an annual rate of increase per capita of between 3.5 and 3.8 percent [Hazlewood 1979]. Succumbing to the worldwide economic recession of the 1970s, this rapid rate of growth could not be sustained, except during 1976-1977 when there were exceptionally high export prices for coffee and tea. In 1979 the GDP grew at a rate of 3.1 percent (in constant prices), and in 1980 at 2.9 percent, both below the nation's population growth rate [Kenya, Ministry of Economic Planning and Development 1980a; International Monetary Fund 1982].

Despite developmental priority having officially been given to agriculture [McChesney 1980], urban-rural developmental inequities have grown, and Kenya's domestic terms of trade have turned increasingly against agriculture and in favor of industry [Gaile 1976; Sharpley 1981]. The nation's development bias toward industry was largely fueled during the 1960s and early 1970s by an expanding agricultural base. However, during the past decade it has been difficult to maintain even existing levels of agricultural production. In recent years Kenya

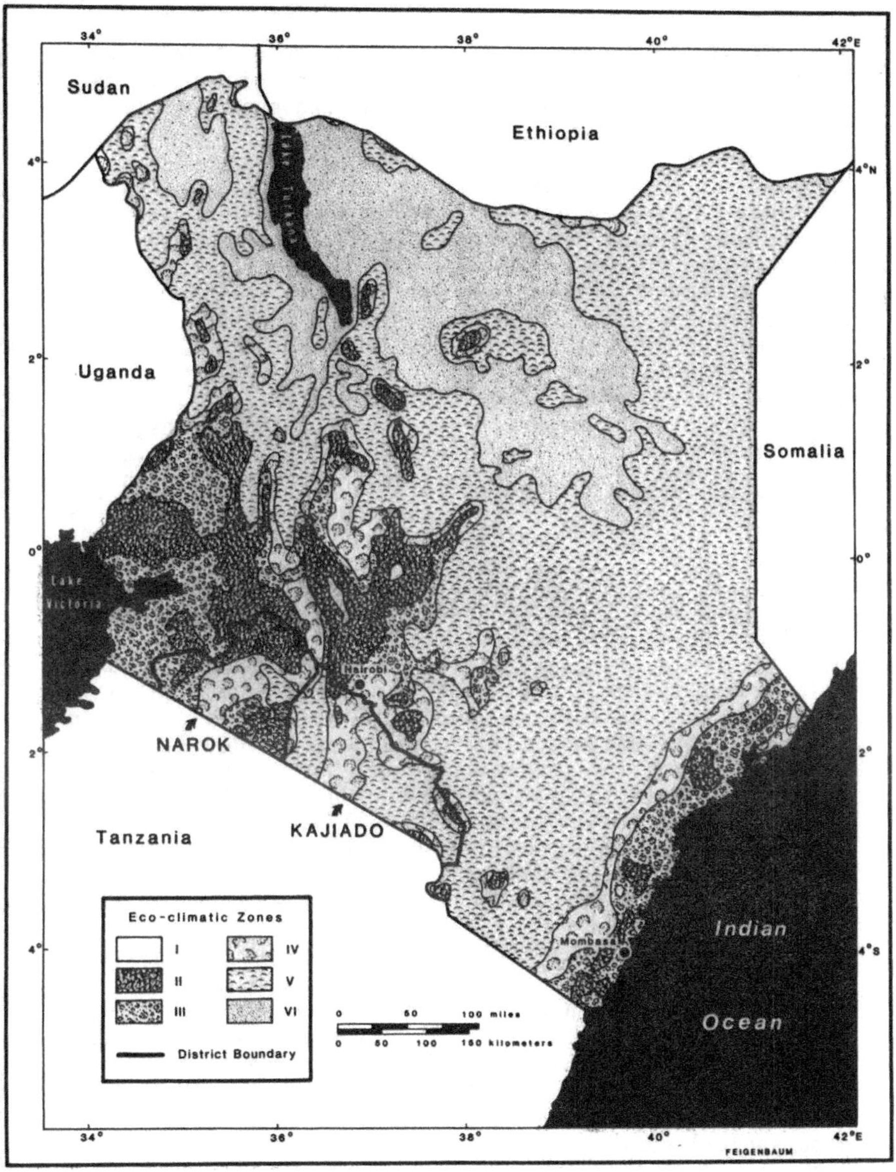

Figure 1.1. Kenya's eco-climatic zones

Source: Pratt and Gwynne (1977), Map 1.

has fallen from agricultural self-sufficiency to become an importer of basic foodstuffs, with 350,000 metric tons (mt) of maize, 118,000 mt of wheat and 13,000 mt of milk powder imported during 1981.

Kenya's economic stagnation, and agricultural decline in particular, is reflected in the recent history of the nation's livestock industry. At the beginning of the 1970s, the World Bank concluded that "in general it can therefore be confidently projected that Kenya will continue to meet the effective demand for beef and will continue to have exportable surpluses of beef in the form of meat or live animals" [IBRD/IDA 1972, p. 28]. By 1980, the marketed supply of carcass beef barely exceeded domestic demand, with supply estimated to be 140,000 mt, and demand 135,000 mt. According to the Kenyan Government's own projections, by 1990 the demand for beef would likely increase to 228,000 mt, while "beef production may in fact decline from present levels by 1990 unless the present production systems, technologies, and policies are changed" [Kenya, Republic of 1980a, p. 9].

A rapidly expanding population, expected to double from 16 to 34 million between 1980 and 2000 [Henin 1981], and increasing incomes underlie the growth in demand for meat, especially beef. Regressing consumption per capita on disposable income, price of beef, a composite price of other meats, and time, Kivunja's [1978] double logrithmic model of the demand for beef yielded an R^2 value of 0.984, with the following elasticities of demand: income, 1.676; price, -0.311; and cross price (composite of mutton, goat meat, and pork prices), 0.132. The increase in demand for meat may have decelerated over the past decade, as the climb in personal disposable incomes has slowed. However, given the population's rapid rate of growth, this respite cannot be expected to last, especially as the country's economy revives. In considering Kenya's potential for meeting the expanding demand, it is instructive to examine the country's ecological setting as well as current trends in the supply and demand for meat.

Eco-climatic potential. Kenya's ecological diversity may be classified into six zones, based on a moisture index [Pratt and Gwynne 1977]. While each of the zones can be described in terms of natural vegetation type and agricultural potential, the defining feature is moisture availability, "the most potent environmental factor in East Africa" [Pratt and Gwynne 1977, p. 41]. The geographical configuration of the zones is shown in Fig 1.1 and relative proportions of the country's total area are listed in Table 1.1.

Eco-climatic Zone I occurs at very high altitudes. A minute portion of Kenya is classified as Zone I land, and the only importance of these places to the nation's livestock industry is indirect as water catchment areas. Eco-climatic Zone II, at elevations below those of Zone I, is characterized by a humid to dry semihumid climate, and its productive potential lies mainly in forestry or intensive agriculture. Similar production possibilities hold for Eco-climatic Zone III, in which a dry subhumid climate is found. Zones II and III constitute Kenya's higher-potential areas, which are principally settled and cultivated by smallholders. Cropping is the dominant activity of these mixed farming systems, with livestock enterprises centering upon milk production both for home and for market. However, beef produced by the dairy stock of these farmers does constitute an important share of the country's meat supply.

The semiarid savannas and woodlands of Eco-climatic Zone IV are marginally cultivable at best, and thus best suited for livestock-based production systems. Generally less than 4 ha are required per stock unit, except where dry seasons exceed six months. (The standard stock unit is assumed to equal a 450 kg steer, having a daily dry matter requirement of about 10 kg.) The actual productivity of livestock operations depends on local plant and soil compositions. Despite low and erratic rainfall, Zone IV areas have increasingly come under cultivation due to population pressures, with misuse of the land frequently leading to environmental degradation [Lynam 1978; Porter 1965].

Eco-climatic Zone V comprises lands more arid than those of Zone IV. Acacia and allied genera are dominant and perennial grasses are widespread but readily disappear when mismanaged. Livestock production is the most appropriate activity, with generally more than 4 ha required per stock unit. More arid still are the very dry lands of Eco-climatic Zone VI where the vegetation is mainly annual grasses. Camels are the livestock species best accommodated in Zone VI, although cattle and small stock (sheep and goats) are also herded. These lower-potential Zones IV, V, and VI comprise Kenya's rangelands, broadly defined as "land carrying natural or semi-natural vegetation which provides a habitat suitable for herds of wild or domestic ungulates" [Pratt and Gywnne 1977, p. 1]. Zones V and VI, in which the cultivable potential is virtually nonexistent, assuming irrigation is not feasible, are the areas principally inhabited by pastoralists.

A country's eco-climatic conditions provide the fundamental opportunity set for its agricultural activities, particularly for systems which utilize low levels of purchased inputs. Even when the marginally cultivable Zone IV lands are excluded, over 70 percent of Kenya is characterized by low livestock carrying capacities (Table 1.2). In other words, Kenya has a large resource base favoring land-extensive livestock production systems. These rangelands have provided the ecological foundation of Kenya's pastoral economies and will be increasingly depended upon to supply livestock for the

Table 1.1. Kenya's eco-climatic proportions

Eco-Climatic Zone	Proportion of Kenya's Area
	----------percent----------
I	1
II	9
III	9
IV	9
V	52
VI	20

Source: Senga [1976].

Table 1.2. Approximate livestock carrying capacities of Kenya's eco-climatic zones

Eco-Climatic Zone	Hectares Required per Livestock Unit
II	0.8
III	1.6
IV	4.0
V	12.0
VI	42.0

Source: From Pratt and Gwynne [1977], Table 8.

national market, given the country's mounting population and shifting patterns of meat production and consumption.

Changing production and consumption patterns. Kenya's estimated 11.5 million cattle and 9 million small stock are raised in a variety of environments and systems. Using the Government of Kenya's broad classification of agricultural producers, suppliers of livestock may be categorized as "large farm," units greater than 8 ha, usually in better-watered areas; "smallholder," units less than 8 ha, also usually in better-watered areas; and "range," commercial and subsistence rangeland producers. With the additional classification of "urban areas" (Nairobi and Mombasa) and "exports" as nonsupplying sources of demand, the country can be partitioned into areas producing more livestock than is consumed (meat surplus) and those consuming more than is produced (meat deficit), with slaughter stock moving accordingly. The estimated shares of beef supply and demand in 1975 following this classification are shown in Table 1.3.

Smallholders, that is, mixed farmers principally located in the more densely populated Zones II and III, are the major producers and consumers of beef. However, the net surplus production by smallholders is dwindling as the populations (food demands) of these areas increase and livestock enterprises give way to the cultivation of crops. The amount of land available for grazing is expected to decline nationally about 3 percent between 1975 and 1990, but this reduction will entail a 14 percent decline in total carrying capacity since the withdrawal of land from grazing will mainly affect the higher-potential areas of the smallholders [Kenya, Republic of 1980a].

While surplus meat production in the smallholder areas is projected to decline, Kenya's urban population is expected to increase eightfold between 1980 and 2000 [Ominde 1979]. Meat consumption per capita in Kenya's urban areas is three times that of smallholder areas, making rapid urban growth all the more significant to the livestock industry's future [Kenya, Ministry of Agriculture 1978]. The decline in higher-potential grazing land, consumption of smallholder production increasingly by themselves, and the rapid expansion of demand in urban areas are prompting Kenya's growing reliance upon the surplus production of its pastoral range areas. Yet, these areas have the lowest commercial offtake rates in the country (Table 1.4), with a large share of the stock slaughtered never entering market channels [FAO/IBRD 1977].

Over the past 20 years, Kenya's production per capita of beef, mutton, and goat meat has remained essentially unchanged at 14 to 15 kg [FAO 1978, 1979, 1982a]. However, simply to maintain this relatively low level of production per capita will be difficult, given the projected doubling of the country's population by the year 2000. Production increases due to larger livestock inventories alone is

Table 1.3 Kenya's beef production and consumption, by type of holding and destination, 1975

	Production		Consumption	
	Quantity	Proportion	Quantity	Proportion
	-1000 mt-	--percent--	-1000 mt-	--percent--
Smallholder areas	102	71	93	65
Large farm areas	22	16	11	8
Range areas	19	13	10	7
Nairobi and Mombasa	--	--	17	12
Exported	--	--	12	8
Total	143	100	143	100

Source: UNDP/FAO [1979], Tables 1 and 2.

Table 1.4. Rates of offtake of cattle herds in Kenya, by type of producer, 1980

Type of Producer	Stock	Rate of Offtake
		----percent-----
Pastoralists	Local zebu	10
Mixed farmers		
Smallholder	Local zebu	13
	Improved zebu	15
	Grade dairy	18
Large farm	Grade dairy	20
	Grade beef	25
Commercial ranchers	Improved zebu	25

Source: Based on Kenya, Republic of [1980a], Table 3.

not a realistic possibility. Kenya's small stock numbers have not increased since the mid-1960s, and while the national cattle herd has grown, the number of cattle per person has declined from 0.74 in 1966 to 0.67 in 1981 (Fig. 1.2). To meet the expanding demand requires higher levels of productivity per animal and per ha, and higher rates of offtake in Kenya's pastoral areas.

Maasailand's Transition

Location and eco-climatic setting. A major pastoral area of livestock production is Kenya's Maasailand, a region which extends over 40,000 sq km (7 percent of Kenya's total land area) along the Tanzanian border Fig. 1.1). Though the Maasai people inhabit northern Tanzania as well, "Maasailand" as used here refers only to that portion of their land located in Kenya. Eco-climatically the region can be roughly divided into dissimilar halves which approximately match the administrative units of Narok District on the more humid western side, and Kajiado District, to the more arid east [Griffiths 1962; Ominde 1968]. The eco-climatic distinctiveness of the two districts is evident in Table 1.5 and Fig. 1.1. Ninety-five percent of Kajiado District falls within the semiarid Zones IV and V, while most of Narok District lies within the higher-potential Zones II and III. Hence, Maasailand has a heterogeneous resource base which overall holds great productive potential. This capacity for expanded livestock production is already beginning to be tapped.

Livestock production. Two-thirds of the beef consumed in Nairobi, as well as most of the capital's mutton and goat meat, comes from Maasailand, attesting to the urban dependency upon pastoral production described above [UNDP/FAO 1980]. However, increased demand at the national (urban) level is not the only reason production must expand. Within Maasailand, deteriorating ecological conditions and an increasingly inadequate subsistence orientation also underscore the need for change. Pressures on land, leading to environmental degradation and diminishing levels of welfare, are undermining the Maasai pastoral mode of production which historically supported a sustaining, if underproductive, economy. Increasing population densities and concomitant socioeconomic disruptions have resulted for the Maasai in what Jacobs describes as "an insidious decline in their total situation," his prognosis being that "indeed, viewed in historical perspective, the immediate future of Maasai pastoralists appears bleak as well as obscure" [1975, pp. 418, 421]. The pessimism expressed by Jacobs conveys the urgency, intraregionally, of overcoming the constraints which limit Maasailand's livestock production.

Under present pastoral livestock production systems, growth prospects for areas such as Maasailand are nil

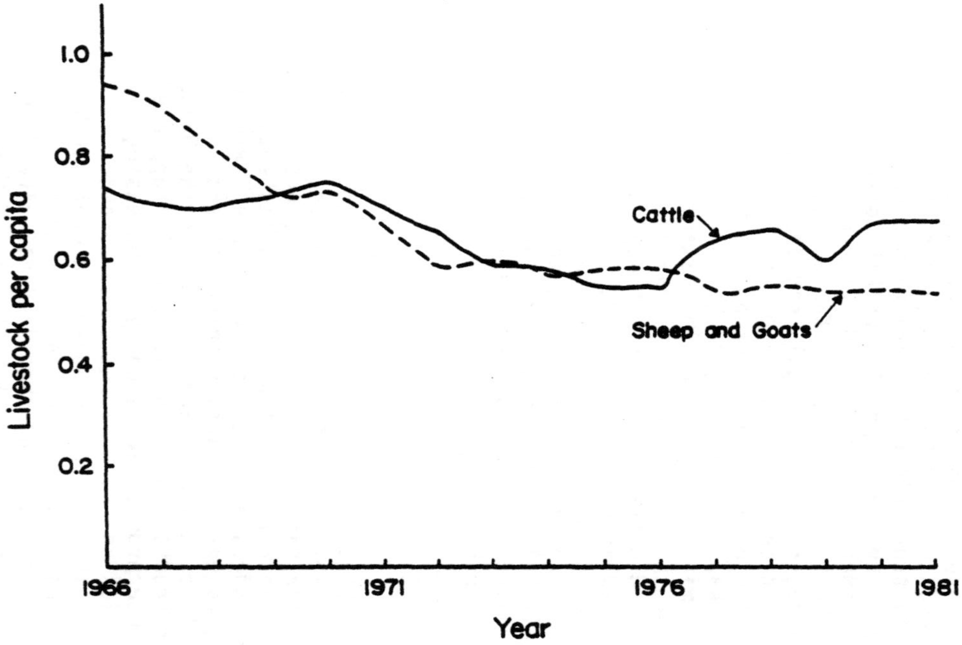

Figure 1.2. Cattle and small stock per capita, Kenya, 1966-1981

Source: FAO (1978, 1979, 1982a).

Table 1.5. Eco-climatic settings of Kajiado and Narok Districts

Eco-Climatic Zone	Proportion of Area	
	Kajiado District	Narok District
	------------percent------------	
II	1	38
III	1	21
IV	36	36
V	62	5

Source: Kenya, Ministry of Economic Planning and Community Affairs [1979a].

[de Montgolfier-Kouevi and Vlavonou 1981]. However, the rangelands offer great scope for increasing livestock production once the transition is made to market oriented production systems [IBRD 1977]. Rising consumption demands and land use pressures make it imperative that the transition process be facilitated by development interventions designed to encourage the shift by Maasai pastoralists to increased market participation.

The transition process is fraught with obstacles and truly involves fundamental change, given the distinctly different objectives and practices which characterize pastoral and market oriented stockraising systems. Market oriented systems depend upon input and output markets, and are generally associated with private access to resources. Ecological limits may be significantly modified through technological innovation. Accumulation of wealth derives from returns exceeding costs of production and growth, from reinvestment of net revenue in the enterprise. Livestock are a means of maximizing net returns to the factors of production. In pastoral societies such as that of the Maasai, resources are communally held and ecological constraints are little modified. Markets in factors and products are appendages to the system, and the selling of an animal is often due to necessity rather than by plan [Dahl and Hjort 1976; Doherty 1979b]. Livestock are repositories of value and the spur to livestock accumulation is underproduction for the market [Ingold 1980; Schneider 1981a].

Despite such differences between the pastoral and market oriented modes of production, the transition from the former system to the latter is occurring. Yet, interventions intended to facilitate the transition in other pastoral areas of Subsaharan Africa have led to controversial conclusions regarding economic and social costs and benefits [Behnke 1982; Frantz 1981; Werbner 1980]. For example, with respect to the development of market oriented ranches in Botswana, Hitchcock challenges measures used to substantiate their desirability over traditional systems. He concludes that "in the long run commercial ranches are likely to be less profitable than the long-standing cattle system in Botswana" [Hitchcock 1980, p. 171], despite empirical evidence supporting the greater productivity of fenced ranching [Rennie et al. 1977]. Livestock development strategies in Maasailand will be pursued with confidence only if the net benefits for Maasai producers as well as for other Kenyans are made clear.

In sum, a common problem of livestock sectors of developing countries is said to be their continual struggle to balance supply and demand [Harrington 1976]. Kenya's balancing of meat production and consumption can be translated into a single objective: expansion of the nation's meat supply in the hope of keeping pace with an inexorably rising demand. Unless conditions constraining livestock production levels in pastoral areas such as Maasailand are overcome, the required supply of livestock will not be forthcoming, and projected meat deficits for the nation will become a reality. Moreover, pastoral peoples like the Maasai will suffer a general decline in welfare as population pressures invalidate traditional production regulating relationships.

For Maasailand's livestock production to expand, the overriding constraints need to be accurately identified and evaluated, in order that effective interventions can be implemented. As the Kenyan Government itself acknowledges, "the greatest challenge that lies ahead now is in developing the appropriate mechanisms for implementing the [country's livestock development] projects" [Kenya, Republic of 1980a, p. 53]. Developing appropriate mechanisms depends ultimately upon a clear understanding of the constraints. Livestock production constraints and development alternatives for Kenya's Maasailand are the issues addressed here.

OBJECTIVES AND APPROACH

The analysis of factors constraining livestock production in Maasailand must be undertaken on several planes. A wide range of influences require consideration, from Kenya's development policies, to tenurial conditions in Maasailand, to the management practices of Maasai pastoralists and ranchers. In recognition of the broad scope of the problem, the following objectives are set forth:

1. Identify and evaluate factors exogenous and endogenous to the production unit which constrain the transition in Maasailand from subsistence to commercially oriented livestock production. This objective entails
   a) consideration of the history of the transition process, in terms of principal components and their interactions,
   b) examination of national policies and activities, in particular price controls, which influence the process,
   c) investigation of the influence of regional economic, sociocultural and tenurial conditions upon the process, and
   d) analysis of (i) producers' production and marketing practices and productivity levels, and (ii) the efficiency of local-level livestock marketing mechanisms.
2. Recommend policy changes which could lead to increased livestock production in Maasailand by facilitating the transition from subsistence to market oriented production.

Given the diverse aspects of the first objective, the method of study must be multi-faceted as well. Central to the methodology followed is a farming systems perspective of the problem [Shaner, Philipp, and Schmehl 1982]. Pastoral livestock systems are comprised of interacting environmental, biological, cultural and economic factors, and an analysis of constraints that does not include a consideration of the interrelationships of these factors is unlikely to lead to useful policy recommendations for increasing production [Fitzhugh and Byington 1978; Jahnke et al. 1978; Levine and Hohenboken 1982; Winrock International 1981]. Within the systems perspective, attention must center upon the production unit (household or ranch), that is, the decision-making body with regard to livestock production and marketing practices and, secondarily, upon the livestock trader, who plays the principal marketing role at the local level. In addition, consideration of national policies and regional socioeconomic circumstances needs to be included in the analysis, if the existing system is to be fully understood [Little 1984]. Constraints to expanded livestock production in Maasailand are therefore examined at both the macro and micro levels.

Using secondary data, national policies and programs, and regional socioeconomic conditions are analyzed as institutions currently delimiting Maasai production possibilities [Parsons and Smelser 1956]. The issues of productivity and efficiency are then addressed using primary data collected from sampled production units and market participants, in explicit acknowledgment of the necessity for a disaggregated approach at the micro level [Dalton 1971]. The focus throughout is upon effects, responses, and activities at the producer level, since as Hageboeck et al. [1979] note, it is at the micro level that one gains insight into the process of change.

The major issues considered at the national level are the impact upon the transition process of the Government of Kenya's livestock and meat pricing policies and the influences of various forms of intervention, including marketing, disease control, breeding, and wildlife management. At the regional level the sociocultural characteristics of the Maasai household economy and the impact of changing tenurial conditions are the principal areas of attention. This examination of the national and regional contexts in which the Maasai producer operates is performed recognizing that the individual producer is not a decision-making isolate and sets the stage for the subsequent micro-level analysis of producer and trader behavior [Bennett 1978; Ferguson n.d.; Godelier 1972; Stanfield 1979].

Three samples of Maasai livestock production units, distinguished by tenurial and ecological characteristics, are the basis for the micro-level analysis of production. Sample selection and survey methods are described in Chapter V and, therefore, are only summarized here. One sample consists of households living on communally owned ranches, called group ranches, located in a semiarid part of Kajiado District. The livestock production and marketing practices of these households typify those of the majority of Maasai producers.

Owners of nine individual (private) ranches comprise the second sample of Maasai producers, also located in Kajiado District. As with the sample of group ranch producers, the objective was to gather information on livestock production and marketing, for the derivation of production budgets.

The third sample, located in a more humid region in western Narok District, includes three small-scale producers operating on communally owned land and two larger-scale private holders. Information on these production units was also obtained by personal interview. The sample site was chosen so that the effects of the region's higher productive potential on the transition process could be examined.

Neither this third sample nor the one composed of private ranchers is large. However, analytical deficiencies due to the samples' small sizes are compensated by the specificity of the data and diversity of sample units included. As Morris [1981] notes with respect to most agricultural research on management and production, simply averaging results usually disguises the most interesting information. The objective here is to reveal the particular circumstances limiting expanded production by examining what producers do, not what they should do [Brown 1979].

Relative efficiency is the issue of central concern in the analysis of local-level livestock trading. Are, for example, marketing inefficiencies preventing increased livestock production in Maasailand? Assessment of market performance in Maasailand centers upon the practices of sampled livestock traders and local-level butchers, with the aim of clarifying the "obscure interaction" [Lele 1975] between traditional and modern marketing institutions. As detailed in Chapter VII, the analysis is approached in terms of the structure-conduct-performance model originally developed in the study of industrial organization, with resulting qualitative inferences regarding performance reconsidered in terms of traders' and butchers' actual costs and returns. Information on sources and destinations of livestock, price margins, trading costs, and slaughter weights were obtained from livestock traders operating in the vicinities of the first and third producer samples described above.

The overall methodology and approach constitute a step-wise consideration of the continuum of constraints preventing expanded livestock production, from an assessment of macro-level issues using secondary data, to an analysis of micro-level production-marketing behavior based on primary data. Policy changes are then proposed, based on the analysis of these data and given the forces of tradition and change which currently permeate

Maasailand [Galaty 1981c; Gruchy 1976]. In sum, principal constraints exogenous and endogenous to the production unit are determined, and recommendations for overcoming these constraints are set forth.

NOTE

[1] In some of the referenced sources, Maasai and Maasailand are spelled as Masai and Masailand. These spellings are left unaltered where they appear in the quoted passages and bibliography.

2
Pastoralism and Interventionism

An analysis of constraints hindering the transition of the Maasai pastoral economy to a market oriented production system requires an understanding of both the basic components of pastoralism and the problems which have been encountered in the implementation of pastoral interventions. The purpose of this chapter is to present the elements and interactions that are keys to this understanding. In the first section the traditional economic objectives of pastoral producers are discussed. Immediate and longer-term subsistence requirements of pastoral households are shown to have led to livestock management strategies in which the market plays but a peripheral role. The counterproductive effects of communal grazing practices in discouraging investment and encouraging the overexploitation of resources are then described. Population pressure is identified as having irremeably disrupted a once viable, if underproductive, livestock system.

The historical development of pastoral interventionism is addressed in terms of the Kenyan, and particularly the Maasai, experience. Pastoral interventions during the colonial period, and the course of rangeland development since independence as reflected in the Kenya Livestock Development Project, are reviewed. The general lack of success of pastoral interventions, in Maasailand and elsewhere, is discussed, and the thesis that interventions need to be designed explicitly to promote the transition to increasingly market oriented production is set forth.

THE PASTORAL MODE OF PRODUCTION

Pastoralism involves the subsistence oriented management of livestock grazing a natural resource base. Ingold [1980] describes it as an appropriative mode of production with pastoralists dependent upon the products of their livestock which, for the Maasai, are principally milk from cattle and meat and fat of small stock. Thus, the pastoralist's assets are of two types, livestock and grazing rights [Barth 1964]. While each pastoral society is unique, common economic characteristics, from the mobility necessitated by a reliance upon natural grasslands to the primacy of livestock accumulation, have resulted in a degree of homogeneity among pastoral peoples [Goldschmidt 1979].

Production is governed by private ownership of livestock and common access to the resource base, institutions which are intrinsic to pastoralism's economic structure. As one authority notes:

> At the lower level is the production unit, formed by a set of humans owning and responsible for managing a set of animals by which they are fed. . . .At the upper level is the resource allocation unit, which is defined more by territory (as a function of resource density and fluctuation probabilities) though it has an associated population. [Dyson-Hudson 1980a, p. 13]

Despite the implied distinction between product appropriation and command over resource use, "ultimately, the production units have autonomy to operate how they like, and to take the consequences of doing so" [Dyson-Hudson 1980a, p. 14].

Pastoralism in Africa cannot be considered apart from its principally semiarid to arid environment. While ecological conditions do not determine a society's form, they do place demands upon its social and economic structure [Bonte 1981; Dahl 1979]. The ecological demands are major for pastoral economies, dependent as they are upon a variable and limited resource base [Kenworthy 1971]. Annual rainfall in pastoral areas is usually below 500 mm, and its incidence, spatially and temporally, is highly unpredictable [Helland 1980a]. The vulnerability of the pastoral mode of production to resource uncertainty is manifested in a cyclical rise and fall in the welfare of pastoral peoples: "The inevitable failure of the rains which brings heavy loss of livestock and grave danger to human lives; the probable better year or two when losses are recouped and the next disaster is prepared for, economically and psychologically" [Gulliver 1978, p. 204].

Notwithstanding their distinct, largely ecologically conditioned mode of production, pastoralists have the same overall objectives as all peoples: material well-being, agreeable social relations, security and power [Ruthenberg 1980a]. Material well-being for the pastoral household depends upon its viability, defined by Helland as "the capability to derive a livelihood from the herd on a sustained basis" [1978, p. 79]. Production must equal or exceed consumption, unless other pastoral households are willing to subsidize it. Given the unpredictability of the resource base, viability can best be assured by owning large numbers of animals, hence the propensity for livestock accumulation. The successful pastoralist is the successful accumulator of livestock [Gulliver 1978].

Viability is the pastoralist's principal concern, while secondary objectives generally contribute--and rarely contradict--strategies for its maintenance [Almagor 1980]. In other words, economic and noneconomic motivations underlying pastoral activities commonly are mutually supportive. The prestige acquired through the increase of one's livestock holdings is one of the more obvious instances of this interrelationship. There may be occasions in which ritual obligations override near-term economic considerations, but even then the social relationships forged or strengthened are likely to be economically beneficial in the long run [Baxter 1975]. As Horowitz [1979] explains, when a pastoral household has more livestock than it can reasonably manage, there are two options: it can sell or consume them, or it can convey them to other persons, thereby converting animals as economic goods into social and political obligations from which there may be future economic returns. In sum, social and economic success depends upon the accumulation of livestock, with exchange and other types of livestock transaction basic to the process of accumulation [Brown n.d.; Dahl 1981; Galaty 1981d; Helland 1980e; Jacobs 1965]. In the words of Goldschmidt:

> In pastoral societies, a successful career is built upon the acquisition of animals which are to serve as the basic source of subsistence, as a measure of personal status, as the basis upon which influence and power in the community is established, and as the means of projecting social status and influence into the future through the acquisition of wives, and hence sons, and also through the establishment of personal obligations that assure protection against the vicissitudes of old age. Thus all the potential satisfactions of social life hinge upon the acquisition of an adequate supply of animals. [1979, p. 22]

Other practices in addition to livestock accumulation help assure viability, from species diversification and the spatial dispersion of livestock, to the keeping of predominantly female herds and flocks. Pastoralists, as portfolio managers conscious of the benefits of owning a diversified set of assets, will frequently keep sheep and goats as well as cattle. Goats, for example, are more efficient converters of plant matter than cattle, are generally more resistant to drought, have shorter reproductive cycles and potentially higher rates of offtake, and possess other traits which strengthen the household's viability during periods of stress [Chemonics International Consulting Division 1977; ILCA 1980c; Jarvis 1974; Livingstone 1977; UNDP/FAO 1979; Wilson 1982]. Herd dispersion is widely practiced as a precaution against local disease outbreak as much as against the vagaries of weather [Dahl 1979; Eidheim and Wilson 1979]. Herds composed primarily of females not only ensure a greater probability of growth and adequate milk supplies, but also more rapid post-drought recovery [Campbell 1979b; Dahl and Hjort 1976; Semenye 1980]. In addition, deferred grazing arrangements and numerous other, often subtle livestock management practices underlie pastoralists' capacity to adaptively utilize their widely variable resource base [Awogbade 1979; Peacock, de Leeuw, and King 1982].

Periods of drought, unpredictable yet inevitable, are the times in which viability strategies are most rigorously tested. For Maasailand, the most serious drought in recent history occurred in 1960-1961, a period of natural disaster prolonged by subsequent torrential rains when the cattle population of Kajiado District, for example, plummeted from about 630,000 in 1960 to 200,000 in 1962. Yet following this disaster the rate at which herd numbers rebounded was almost as astonishing as their precipitous decline. Cattle numbers are estimated to have doubled in three years to 429,000 in 1965, and recovered to pre-drought levels in seven years [Meadows and White 1979]. The rapid rate of recovery may be largely attributed to the good grazing conditions which prevailed in 1963 and 1964, but pastoral herd structures and Maasai management practices, especially the preferential care given young females during the drought, undoubtedly accelerated recovery.

Droughts accentuate the fluctuating interaction between a pastoral population and its natural environment, a relationship dependent upon the size of the human population, the size and structure of herds and flocks, and the capacity of the environment to provide adequate water and forage [Western 1982]. As long as fluctuations in herd size do not exceed an ecologically set upper limit or fall below a lower limit set by subsistence requirements, viability is maintained. Notably, however, the pastoral household exerts little direct control over this balance between livestock numbers and available grazing. In place of control one finds reliance upon processes of anticipation and response typical of preindustrial societies, that is, adjustment mechanisms which are flexible and easily abandoned, low in capital requirements, and which depend upon action only by small groups [White 1974].

The characteristic of a minimal degree of control over productive resources extends beyond environmental factors, as suggested by pastoralists' low levels of investment in livestock and labor, system components over which they have some command [Jarvis 1980]. It has been posited that a low investment level is the dominant trait of pastoral economies [Schneider 1979]. This reliance upon adjustment processes rather than control measures is a successful strategy so long as there is an adequate communal resource domain. Thus, human and animal population densities are the crucial parameters. If a region's pastoral population remains relatively static, or its growth can be accommodated by territorial expansion, production can be sustained. But when there is population growth without territorial expansion, or worse, a reduction in territory as has taken place in Maasailand, the pastoral mode of production becomes destructive to both its environment and society.

In the end, persistence through time has been the singular achievement of the pastoral strategy throughout Subsaharan Africa [Dyson-Hudson 1980b]. Given the productive variability of the resource base and the negligible control pastoral peoples have been able to exert over it, this achievement is not trivial. But persistence through time is no longer sufficient given the production demands increasingly placed upon pastoral regions of the continent. Direct and opportunity costs of the pastoral production mode as population pressures mount, in misuse of range resources and livestock production foregone, are becoming greater than Kenya and the other nations of Subsaharan Africa can afford to bear.

POPULATION PRESSURES

As explained in Chapter I, a rapidly increasing population density--for Kenya in general and Maasailand in particular--will inevitably necessitate changes in production practices. Traditional Maasai producers will face an ever greater struggle simply providing for their own households, let alone expanding meat production to meet the demands of the larger Kenyan society. Because the population and land pressure issues are of utmost significance to Maasailand's current economy and future options, they are discussed here in greater detail.

Ecological Balance

Pastoralism, as depicted in the preceding section, is abstracted from the processes of change set in motion by increasing population densities [Konczacki 1978; Ominde 1971]. As described, in the absence of technology for increasing productivity, the environment sets a limit on herd size, and ultimately on the size of the subsistent human population [Swift 1975]. This limit is determined by the land's carrying capacity, a complex and highly variable property, influenced by climate, soil and plant compositions, livestock management practices, and other factors. If a pastoral population grows but cannot expand spatially, the land will come under increased pressure, until a "critical population density" is reached [Helland 1980a]. The population exceeds this limit only to the detriment of the resource base, that is, in reducing the area's primary productivity.

Numerous studies [Baker 1975; Crotty 1980; Hjort 1981; Riney 1979; Stiles 1981; Unesco/UNEP/FAO 1979] have contributed to the debate on whether a balance between populations and resources has traditionally characterized pastoral economies, and proponents on both sides have had a focus for disagreement in the history of the Maasai. On the one hand are sources which identify the past as a time of equilibrium: "Until the beginning of the present century there had been a balance between land resources and livestock/wildlife numbers in the range areas" [Ayuko 1980, p. 163]. Opinions to the opposite are as strongly held: "The Maasai never achieved a stable balance with the resources of their environment in any one area" [Talbot 1972, p. 703]. The controversy continues in terms of the extent of current resource deterioration in Maasailand. Studies on the one side identify all the forms of habitat destruction usually associated with pastoralism within the region's borders, and associate them with the declining self-sufficiency of the pastoral Maasai [Cossins 1980; Glover and Gwynne 1961]. Other reports contradict these accounts, at least as to the seriousness of the problem. For example, in his investigation of range conditions in Tanzania's Maasailand in the late 1970s, Jacobs [1978] found that rangeland conditions were comparable to what they had been 20 years earlier, despite poor rainfall during the period 1970-1976. His conclusions were that "claims of extensive or excessive overgrazing to the degradation of pasture in Maasailand are, in general, both unsubstantiated and wildly exaggerated" [Jacobs 1978, p. 4]. Clearly, the validity of such statements depends upon the locality investigated and the time frame. Ecological forces and historical patterns of interaction that have shaped the Maasailand habitat are complex [Jacobs 1980b]; some areas of Maasailand have suffered more serious degradation of resources than others, and the regenerative powers of the land are far from uniform.

While the divergence of opinion regarding the extent of actual land deterioration remains, there is little argument that population pressures are presently creating a state of increasing imbalance for the pastoral Maasai. They may be among the wealthiest of cattle-owning peoples in Africa [Jacobs 1975], but this claim to well-being is tenuous when one considers Maasailand's declining livestock numbers per person and its shrinking pastoral resource base, as described below. This overexploitation of the range due to rising population pressures has been fostered by the pastoral institution of common access to range resources.

Communal Resource Use

Communal grazing practices have been the norm in many parts of the world where livestock have been grazed extensively. As noted by Crotty, "communal grazing of the many small, individually owned herds in the group combined the economically most productive and the politically most stable methods of exploiting grazing lands by precapitalist pastoralists" [1980, p. 120]. Invariably, however, there has eventually emerged an accompanying history of resource deterioration.

Theoretically, the problem is one of divergence between private and social costs and benefits [Stryker 1984]. The individual pastoral household reaps the full benefit of additional animals, while the cost in pasture, water and eventual range degradation due to overgrazing is shared by all households. The individual pastoralists behave rationally to maximize their expected utilities, thereby assuring the collective failure of all: the so-called tragedy of the commons.

> Adding together the component partial utilities, the rational herdman concludes that the only sensible course for him to pursue is to add another animal to his herd. And another; and another. . . .But this is the conclusion reached by each and every rational herdman sharing a commons. Therein is the tragedy. Each man is locked into a system that compels him to increase his herd without limit--in a world that is limited. Ruin is the destination toward which all men rush, each pursuing his own best interest in a society that believes in the freedom of the commons. Freedom in a commons brings ruin to all. [Hardin 1968, p. 1244]

Approaching the common property problem from a different perspective, Runge [1981] argues that it is the uncertainty of the individual pastoralist regarding the actions of others that results in the overexploitation of common property resources. In his opinion, the separability assumption entailed by a model of each pastoralist strictly pursuing self-interests is invalid. Rather, individual choices by pastoralists need to be viewed as interdependent, with the uncertainty of strategic interdependence providing the rationale for institutions coordinating choice [Runge 1982]. By this theory, grazing decisions are based on the expected decisions of others, and "overgrazing results from the inability of interdependent individuals to coordinate their actions" [Runge 1981, p. 604].

In unqualified form, these analyses of the commons problem assume unrestricted access by the individual pastoralist to communal resources, without limitation on livestock numbers or on timing/duration of resource use. However, a "common property resource," when distinguished as an analytical classification from a "public good," implies historically derived institutional arrangements which govern resource use [Siebert 1981]. Regulating mechanisms have developed within most pastoral societies that militate against the overexploitation of the resource base, and grazing and watering rights are frequently circumscribed by social law and custom [Sandford 1983]. Still, the effectiveness of institutionalized restrictions erodes as an area's critical population density is exceeded. Voluntary compliance in the absence of outside enforcement becomes less likely [Palmquist and Pasour, Jr. 1982]. Whether the common property problem is perceived as an assurance problem derived from informational deficiencies, attributed to a divergence between social and private costs and benefits, or in fact results from a combination of these and other factors, the outcome is plain: as population densities increase, grazing pressures are unlikely to be held at optimal levels under communal land use systems. The impact on production is likely to be considerable. Jarvis [1980] estimates production losses due to communal tenure can equal as much as one-third of what could be produced under individual tenure.

Encapsulation and Encroachment

Communal tenure allows, even encourages, a welfare loss, but the magnitude of the problem is dependent upon the intensity of land use pressures, principally a function of pastoral population growth. In areas like Maasailand, political encapsulation and encroachment by nonpastoral peoples have also significantly affected population densities.

Loss of access to grazing resources during colonial rule, usually the better-watered areas, fundamentally redefined production possibilities for many pastoral peoples. As stated by Salzman, "in an understanding of the circumstances of pastoral peoples today, the political fact of encapsulation is foremost" [1981a, p. 131]. In the case of the Maasai, they dominated an area in Kenya and northern Tanzania that extended 1,100 km north to south and up to 300 km east to west during the latter half of the nineteenth century. The British appropriated much of the higher-potential dry season land in Kenya for colonial settlement, and the Maasai were restricted through a series of "agreements" first to southern and northern reserves, and then to an expanded southern reserve alone [Spencer 1983; Tignor 1976]. By 1911, Kenya's Maasailand had been reduced essentially to the present boundaries of Kajiado and Narok Districts. Since then, the process of encapsulation has proceeded, though under more conciliatory circumstances, in the establishment of game parks and reserves within Maasailand [Campbell 1977]. The process of encapsulation has resulted in a situation similar to that of the American Indian in the western United States [Simpson 1968].

In addition to politically legitimized losses of territory, the resource base of the pastoral Maasai is shrinking as population pressures in highland areas compel cultivators to migrate to previously unsettled areas in Maasailand [Bernard and Thom 1981]. Tribal territorial exclusivity, still a potent political element in most of Subsaharan Africa, has limited incursions into Maasailand to a degree, but the general trend is one of in-migration and settlement by agropastoral groups. The proportion of Kajiado District's total population counted as Maasai decreased from 78 percent to 63 percent between 1962 and 1980, despite an increase of the Maasai population by 75 percent [White and Meadows 1981].

In agropastoral production systems cropping is usually the basic activity, with surplus income invested in livestock. Often the newly farmed areas are only marginally cultivable Zone IV and Zone V lands, where cropping is of doubtful ecological soundness [Marimi 1979]. The country's policies regarding such agricultural expansion have been described as revolving around the question of "how many people can be maintained adequately in these marginal areas without the necessity of large-scale famine relief in the inevitable low rainfall seasons" [Lynam 1978, p. 64]. Not only are the more productive range areas lost to cultivation, but grazing pressures are intensified by agropastoralists' herds and flocks. In addition to direct displacement from grazing lands, the hydrologic impact of cultivators moving into water catchment regions has hastened the impoverishment of Maasailand's resource base [Ambrose 1980]. Even when yield risks are taken into consideration, cropping of Zone IV pastoral areas is a productively rational course for the agropastoralist [Jahnke 1982], and therefore is a trend likely to continue.

For Kajiado District, the loss of dry season grazing areas as agropastoralists settle marginal lands, not to mention the impact of the expanding urban environs of Nairobi, is an ongoing process measurably reducing the pastoral resource base ["Kajiado, District Environmental Assessment Report" 1980]. In Narok District, with its higher agricultural potential, encroachment is taking place on a yet larger scale. The decline in the district's resource base, in many areas to levels below that required to support subsistence pastoralism, is a trend greatly aggravated by the increased cultivation of former dry season grazing areas [Cossins 1980]. Political encapsulation diminished the grazing resources traditionally depended upon by the Maasai in the past; presently, encroachment by agropastoralists is furthering the loss. As the pastoral resource base narrows, land use pressures are compounded by a rapidly expanding Maasai population.

Maasai Population Growth

The desire of every Maasai pastoralist for children is economically rational, considering the labor demands of pastoral livestock production and the reliance upon one's offspring for old-age security. But, having children is imbued with a cultural significance beyond the purely economic. As observed by Jacobs,

> in spite of the importance which Maasai attach to their cattle, they regard the possession of children, especially sons, as a more significant social value. Hence, a "wealthy man" (olkarsis) is one who has both many cattle and many children; a man with many cattle but few or no children is called tetia (pl. tetiain) and is accorded no special status at all, being equated with a "poor man" (olaisinani) who has neither many cattle nor children. [1965, p. 332]

Population growth rates of pastoral peoples have been found to be generally lower than those of cultivators [Campbell 1979a; Swift 1982]. A prolonged period of breastfeeding among the Maasai, for up to 3 1/2 years, during which time intercourse is considered harmful to the feeding child, is one practice which has resulted in the spacing of children [de Souza 1980]. Also, the fact that traditionally a man cannot marry until he has attained senior warriorhood has tended to restrain population growth rates.

The trend, however, is toward younger and larger families as both Maasai men and women marry at an earlier age. De Souza [1980] attributes the younger marriages to a general lowering of the age at which a young man is initiated into warriorhood, and even, in places, the dissolution of this ceremony; the spread of school attendence and its disruptive effects regarding traditional social practices; and a decline in many pastoralists' welfare, inducing fathers to have daughters marry at a young age so that the brideprice may be received sooner than later. When improved health care which is now available is considered in addition to these social changes, it is not surprising to find an accelerating population growth rate among the Maasai.

Human and livestock population densities and ratios for Kajiado and Narok Districts are shown in Table 2.1. The livestock numbers, in particular, while not unrealistic, should be recognized as only approximations since animal censuses are highly prone to inaccuracies. Livestock holdings per capita for other pastoral peoples of Kenya are presented in Table 2.2 for comparison. Current population densities depict existing land use pressures, but it is the trends over time which substantiate the increasing gravity of Maasailand's population problems. The results of two sets of projections now described indicate that subsistence needs soon will be greater than the capacity of the resource base, as utilized in the pastoral mode of production, to provide for them.

In one study [Kenya, Ministry of Economic Planning and Community Affairs 1979a], the 1969 census is used to project pastoralist and farmer populations in Kajiado and Narok District (as well as populations in other pastoral areas) to the year 2000. These populations, expressed in adult equivalents, are shown in Table 2.3. In Table 2.4 the maximum numbers of livestock units environmentally supportable, given the eco-climatic conditions of the two districts, are calculated. Using this information, the maximum numbers of pastoralists supportable, assuming various land use restrictions and a minimum requirement for subsistence of 3.5 standard livestock units per adult equivalent, are projected, along with the years by which the pastoral population of Maasailand can be expected to exceed these limits (Table 2.5). For example, assuming pastoralists of Kajiado District are prevented from grazing the cultivable Zone II and Zone III lands, they will exceed the carrying capacity of the remaining range by 1993.

The second study cited ["Kajiado, District Environmental Assessment Report" 1980] employs the same methodology, and though only pertaining to Kajiado District, a wider range of conditions are investigated. Various scenarios featuring different Livestock Unit/ Adult Equivalent (SSU/Adult) ratios, populations growth rates, levels of technology, and land use restrictions, are considered, as summarized in Table 2.6. For example, if it is assumed that (i) the population growth rate is 2.5 percent, (ii) the stock per capita ratio is 3.5 SSU/Adult, (iii) range productivity increases by 25

Table 2.1. Human and livestock populations and population densities, Kajiado and Narok Districts, circa 1980

Item	Kajiado District	Narok District
Area in sq km[a,d]	19,605	16,115
Population		
Human[a]	149,005	210,306
Cattle	600,000-650,000[b]	602,000[c]
Small stock	701,000[b]	1,254,000[c]
Population per ha		
Human	0.076	0.131
Cattle	0.306-0.332	0.374
Small stock	0.358	0.778
Cattle per capita	4.0-4.4	2.9
Small stock per capita	4.7	6.0

Source: [a]Kenya, Central Bureau of Statistics, 1979 census.

[b]White and Meadows [1981].

[c]Kenya, Ministry of Economic Planning and Community Affairs [1971a], Appendix 3.

[d]Game parks not included.

Table 2.2. Livestock per capita for various pastoral peoples of Kenya, 1980

District	Pastoral People	Livestock		
		Cattle	Small Stock	Camels
		----------head per person----------		
Marsabit	Borona, Rendille, Gabra	5.2	13.5	3.9
Isiolo	Borona, Somali	2.7	9.5	3.6
Wajir	Borona, Somali	1.4	0.65	0.6
Turkana	Turkana	1.02	17.7	1.06
Garissa	Somali	5.9	1.2	0.15
Mandera	Somali	1.05	1.9	1.2
Samburu	Samburu, Rendille	5.2	3.7	0.13

Source: Based on Helland [1980a], Table 1.

Table 2.3. Projected growth of pastoral and farming populations, Kajiado and Narok Districts, 1980-2000

	Population[a]			
	Kajiado District		Narok District	
Year	Pastoral	Farming	Pastoral	Farming
	adult equivalents			
1980	62,927	30,992	88,738	47,988
1990	78,225	43,717	110,311	67,486
2000	97,242	61,667	137,129	88,738

Source: Kenya, Ministry of Economic Planning and Community Affairs [1979a], Appendix X, Table 3.

[a] Expressed in adult equivalents (AE). One adult = 1 AE; one child = 0.67 AE. Populations are projected from 1969 census data at annual rates of 2.2 percent for pastoralists and 3.5 percent for farmers.

Table 2.4. Maximum number of standard livestock units supportable, by eco-climatic zone, Kajiado and Narok Districts

	Livestock Units[a]	
Eco-Climatic Zone[b]	Kajiado District	Narok District
II	24,699	836,930
III	13,125	238,125
IV	186,000	161,500
V	105,900	6,583
Total	329,724	1,243,138

Source: Kenya, Ministry of Economic Planning and Community Affairs [1979a], Appendix X, Table 6.

[a]One livestock unit = 450 kg liveweight, about two head of local zebu cattle.

[b]Assumed carrying capacities as indicated in Table 1.2, and proportions of districts as indicated in Table 1.5.

Table 2.5. Maximum number of pastoralists supportable under different land restrictions, and year by which the pastoral carrying capacity will be exceeded, Kajiado and Narok Districts

Land Restriction	Kajiado District		Narok District	
	Max. Supportable Pastoral Population[a]	Year Pastoral Pop. Will Exceed Capacity	Max. Supportable Pastoral Population[a]	Year Pastoral Pop. Will Exceed Capacity
	-------A.E.--------	-------year---------	-------A.E.--------	-------year---------
All land available to pastoralists	94,207	1999	355,182	2044
Pastoralists restricted from Zone II	87,160	1995	116,059	1993
Pastoralists restricted from Zones II and III	83,400	1993	48,023	Before 1969

Source: Kenya, Ministry of Economic Planning and Community Affairs [1979a], Appendix X, Tables 7 and 8.

[a]Based upon 3.5 livestock units per adult equivalent (A.E.).

Table 2.6. Year by which the pastoral carrying capacity of Kajiado District will be exceeded under different stocking rates, levels of technology, land use, and population growth rates

		Land Availability and Level of Technology					
		All Land Available			Zone II Land Not Available		
Stocking Ratio	Rate of Population Growth	Current Carrying Capacity	Current Capacity + 25%	Current Capacity + 50%	Current Carrying Capacity	Current Capacity + 25%	Current Capacity + 50%
standard stock units -per adult equivalent-	-percent--	-----------	-----------	-----------	year	-----------	-----------
3.5	2.2	1997	2008	2017	1994	2004	2013
	2.5	1996	2006	2014	1994	2003	2010
	3.0	1995	2003	2010	1993	2001	2007
2.0	2.2	2024	2035	2043	2020	2031	2040
	2.5	2020	2030	2037	2018	2027	2034
	3.0	2017	2025	2031	2015	2022	2028
1.0	2.2	2055	2066	2074	2052	2062	2070
	2.5	2048	2058	2066	2046	2055	2062
	3.0	2040	2049	2055	2038	2045	2051

Table 2.6. continued

Stocking Ratio	Rate of Population Growth	Land Availability and Level of Technology					
		Zone II and III Land Not Available			Zone II, III, and One-Fourth of IV Not Available		
		Current Carrying Capacity	Current Capacity + 25%	Current Capacity + 50%	Current Carrying Capacity	Current Capacity + 25%	Current Capacity + 50%
standard stock units -per adult equivalent-	-percent--	----------	----------	----------	year----------	----------	----------
3.5	2.2	1993	2002	2011	1985	1995	2003
	2.5	1993	2001	2009	1984	1994	2002
	3.0	1992	1999	2006	1984	1994	2000
2.0	2.2	2018	2029	2038	2010	2021	2030
	2.5	2016	2025	2033	2009	2018	2026
	3.0	2014	2021	2027	2008	2016	2021
1.0	2.2	2050	2060	2068	2042	2053	2061
	2.5	2045	2053	2061	2037	2046	2054
	3.0	2037	2044	2050	2031	2038	2044

Source: "Kajiado, District Environmental Assessment Report" [1980], Table 5.1(b).

percent, and (iv) pastoralists retain access to all but the higher-potential cultivable lands, the district's carrying capacity will be exceeded by the year 2001. The projections made in both of these studies lead to the same conclusion: The pastoral mode of production will not be a feasible economic alternative for Maasailand's future populations, a situation apparent in Table 2.7 and Fig. 2.1 which depict the trend of declining numbers of livestock per pastoralist. Between the years 1970 and 2000, numbers of livestock units per pastoral Maasai adult equivalent will have been reduced nearly one-half, assuming higher-potential regions are devoted to the cultivation of crops.

Exemplifying pastoral responses to rising population pressures and a deteriorating rangeland is the increasing prevalence of small stock in the Kaputiei section of Maasailand. Sheep and goats constituted 45.0 ± 10.9 percent of total household livestock units in 1977, having increased from 14.6 ± 7.5 percent in 1967 [Njoka 1983]. But adjustment processes such as this only postpone the need for eventual fundamental change in the mode of production. In the words of Njoka, "one wonders how long the degraded environment can continue to sustain an increasing population of the small stock" [1983, p. 232].

A description of populations and resource use would be incomplete without mention of the region's wildlife, since large populations of predators and herbivores significantly influence the pastoral economy of the Maasai. Predators are a major source of livestock losses, and specific examples involving sampled producers are discussed in later chapters. Concerning wild herbivores, it was estimated in an aerial survey conducted from November, 1969, to March, 1970, that they composed 37 percent of total herbivore biomass in Narok District, and 17 percent in Kajiado District [Watson 1975]. With the national ban on game hunting since 1977, and present livestock populations in Maasailand approximately the same as at the time of Watson's aerial survey, it is likely that the wild/domesticated herbivore ratio has risen, thus placing additional pressures upon grazing resources.

Summary

Population densities unsustainable by the pastoral mode of production are bringing environmental destruction and economic decline to Maasailand and other pastoral areas [Livingstone 1979]. Growing populations unable to expand territorially in association with communal tenure are causing population densities to exceed carrying capacities in some areas, leading to degradation of resources. The in-migration of external populations has compounded the problem, both heightening grazing pressure directly, in pastoralists' dry season grazing areas, and in reducing the productivity of the resource base

Table 2.7. Livestock units per pastoral adult equivalent, Maasailand, 1970-2000

Year	Pastoral Population	Livestock Units per Adult Equivalent[a]
	--adult equivalents--	
1970	122,000	3.8
1980	151,665	3.0
1990	188,536	2.4
2000	234,371	2.0

Source: Based on data from Appendix Tables 2.3 and 2.4 , and Kenya, Ministry of Economic Planning and Community Affairs [1979b].

[a]Assuming only Zones IV and V are used for pastoral livestock production, and that the carrying capacity of these lands totals 459,983 livestock units.

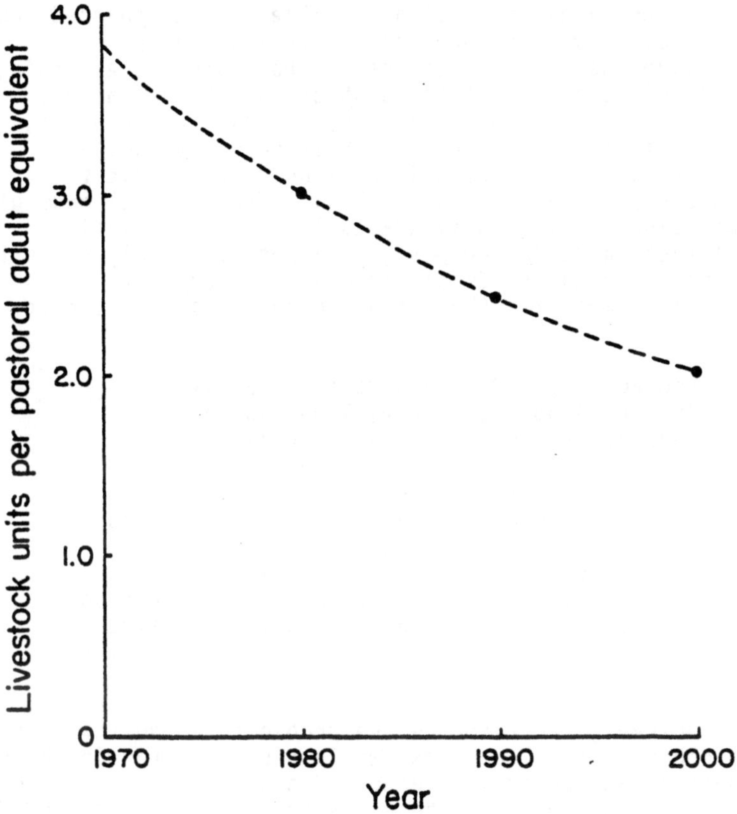

Figure 2.1. Livestock units per pastoral adult equivalent, Maasailand, 1970-2000

Source: Table 2.7.

indirectly through the cultivation of catchment areas. Controversy continues regarding the extent to which resource degradation has occurred, but there is little argument that it is taking place.

As critical population densities are approached and exceeded, adjustment processes have lost their effectiveness. Social structures have undergone strain, and traditional institutions which once helped assure pastoral viability have been undermined. For example, institutions for sharing risk have been weakened and pastoralists have become increasingly vulnerable to climatic uncertainty [Swift 1979]. The pastoral mode of production, ecologically adaptable and productively self-sustaining in the past, will not adequately serve the altered economic needs of the Maasai or of Kenya. As Campbell summarizes the situation for Kajiado District:

> Probably by 1990 and almost certainly by 2000 Kajiado District will be unable to support its population at a subsistence level at current levels of technology. If a major deterioration in the resource base, high unemployment, landlessness and/or livestocklessness and high rates of rural-urban migration are to be avoided, then alternative patterns of resource exploitation, land use and employment will have to evolve before the turn of the century. [1979a, p. 15]

However, the resource use problems facing the Maasai and other pastoral peoples, while unprecedented in their magnitude and urgency, are not without antecedence. In the following section, their historical basis and the evolution of pastoral interventionism in Kenya are examined.

INTERVENTIONISM IN KENYA

Kenya's livestock interventions are addressed in Chapter III. However, to provide this discussion of interventionism with a point of reference, a brief account is given of relevant events in Kenya's past and the country's present framework for development of its rangelands. The scope and relative success of colonial interventions in pastoral regions, Maasailand in particular, are discussed. The objectives and accomplishments of the Kenya Livestock Development Project are then presented.

Colonial Kenya

As with all countries of Subsaharan Africa, Kenya's past can be arbitrarily divided into three eras: the time before colonization by Europeans, an interval of colonial rule, and the period commencing with independence. Each of the three eras spans a successively shorter period of time, from thousands of years, to about 75, to a relatively brief 21 years. In the familiar imperial pattern, British involvement in East Africa was prompted by political and economic opportunism. Control of the Lake Victoria region was the objective, and the Imperial British East Africa Company was granted a concession in 1888 to establish and protect the trade channels from the coast to this area. By 1894-95 the British Government had assumed direct administrative responsibility in East Africa, ushering in Kenya's colonial era [Oliver and Atmore 1981].

Between 1896 and 1901, a railway line was constructed from Mombasa to Lake Victoria. The railway opened up a region of high agricultural potential in what was then the East African Protectorate. Ecologically conducive to European settlement, official encouragement for settlers to establish agricultural holdings in this region soon followed. With the influx of settlers, Kenya's agricultural development was characterized above all by the division of the land between Europeans and Africans. Africans were displaced and subsequently prohibited from acquiring land in what came to be known as the White Highlands, primarily higher-potential lands that eventually totaled about 3.1 million ha. Estates mainly producing wheat, coffee, tea, and maize were established, with the displaced African populations providing a surrounding pool of agricultural labor [Odingo 1971; Stewart 1981]. Parenthetically, settlers' livestock enterprises tended to lag due to problems of disease control and pasture development, and were usually a sideline in farming operations [Hinga and Heyer 1976].

From the beginning, the welfare of settlers was the sole concern of the colonial government:

> Agricultural development policy, that is to say provision of loans, administrative services, roads, market outlets, and research, was up to the mid-thirties almost exclusively, and up to the mid-fifties largely, a matter of promoting and firmly establishing the White Highlands. [Ruthenberg 1966, p. 4]

But the benefits settlers derived were bought very much as the expense of African agriculture and welfare [Smith 1976]. By the 1930s, the adverse effects of neglect of African areas could no longer be ignored. Population pressures and the concomitant deterioration of African lands neighboring the White Highlands led to the appointment of a Kenya Land Commission in 1933. Recommendations for the establishment of planned settlement schemes in African areas were set forth, but it was not until after World War II that these plans were finally enacted. Meanwhile during the war, the contrast between benefits enjoyed by the settlers and the economic difficulties experienced by African producers increased [Spencer 1980].

In 1945, the African Land Development program (ALDEV) was organized, at the same time that the Kenya Ten Year Development Plan (1946-1955), the first of its kind, was about to be initiated. In the Plan, overpopulation and overstocking, which had worsened during the war years, were identified as the salient problems in the African areas. Marginal, nonpastoral African lands which had continued to decline in fertility due to "permanent" bush-fallow practices were the main worry, though rangeland degradation was also beginning to generate official concern. ALDEV was established primarily to oversee the settlement and reconditioning of regions considered cultivable.

Pastoral peoples, even ones such as the Maasai from whom large tracts of land had been appropriated, were in general cushioned by distance against the direct effects of European settlement, and largely left to themselves by the colonial government. Existing tribal boundaries of pastoralists were formalized for administrative control, but otherwise the British presence amounted to little more than pacification. Brown, speculating upon the possible origins of this hands-off attitude, suggests that the pastoral areas were intentionally neglected

> largely because of their inherently poor productivity, and also because the people inhabiting them were generally politically inactive, as compared with agriculturalists. Provided these people did not create an acute problem, in terms of famine, law and order, stock disease, or any combination of these, they could be largely left to themselves, and they were. Indeed, to a degree, there was a deliberate intent to leave the pastoral peoples in their pristine state, not only because rapid change was apparently not necessary for them, but also because many officials admired the warlike, independent, self-sufficient pastoralist more than a trouser-wearing, half-educated, politically minded agriculturalist. [n.d., p. 18]

The minimal level of administrative engagement of pastoral peoples can also be attributed to social and cultural factors which restricted avenues of access. Whereas the social structure of settled African cultivators was often susceptible to colonial manipulation, among the Maasai there was no strong collaborative elite [Tignor 1976]. However, governmental involvement in Maasailand and other pastoral areas became imperative in the face of deteriorating land conditions. Action was spurred not only by the increasing severity of land degradation, but also by a shift in the official attitude toward development during the 1950s. Mismanagement of resources, not simply overpopulation, came to be identified as the source of land use problems [Migot-Adholla and Little 1981].

The change in perception of land use problems was formalized in the Swynnerton Plan, a policy paper on agricultural development in the African areas compiled in 1954. The major thrust of this paper with respect to the nonpastoral areas called for agricultural intensification and land tenure reform, with the consolidation of fragmented holdings and their registration considered necessary preconditions for development [Memon 1981]. For the pastoral areas, the Swynnerton Plan proposed a strategy which called for limiting numbers of livestock to presumed carrying capacities. Guided by this principle, the government launched over 40 major schemes involving stock limitation and grazing management, livestock marketing, water development, and tsetse fly eradication [Migot-Adholla and Little 1981].

Grazing schemes were planned, not according to where they would yield the quickest return but rather mainly where they were most urgently required, with the following objectives and conditions [Ruthenberg 1966]:

- Demarcation of a certain area as a grazing scheme.
- Adjustment of stock to the carrying capacity of the land within the given area.
- Organization of a culling program and stock disposal.
- Castration of scrub bulls.
- Introduction of rotation grazing--simple alternation in most, and regular rotation in some of the better schemes.
- Construction of dips and the provision of veterinary supervision.
- Provision of water through the construction of dams and wells.
- Extension of grazing through bush clearing and tsetse eradication where necessary.
- Provision of loans for the investments mentioned above.
- Payment of a fee per head of cattle per year, to service the loans.

- Supervision of the management rules by grazing guards and a government officer.

Three grazing schemes established in Maasailand were the Ilkisongo (0.5 million ha), the Loodokilani (0.8 million ha), and the Matapato (0.4 million ha) [Ayuko 1980]. Ilkisongo was subdivided into three clan areas, and Loodokilani and Matapato, into grazing blocks. A simple system of management based on the movement of livestock between wet and dry season grazing areas was proposed for each area or block. The regular sale of livestock, selective breeding and culling, provision of water and veterinary services, and most of the other prescribed guidelines listed above were included in the planning. However, in the end the three schemes were not successful.

Failure was due in part to problems of enforcing territorial divisions which conflicted with those traditionally employed by the Maasai [Jacobs 1963]. More significantly, from the beginning pastoralists were required to reduce livestock numbers, since the schemes were located in areas considered overgrazed. Destocking was logically resisted by pastoralists already living at subsistent levels. Moreover, the greatest absolute reductions in stock numbers were required of pastoralists with the largest herds. As these individuals were often community leaders, the schemes were therefore opposed by the most influential members of the society ["Kajiado, District Environmental Assessment Report" 1980].

A Kajiado District demonstration ranch begun in 1946 at Konza experienced a similar fate. It was a 8,870 ha unit, intended to demonstrate to the Maasai the value of limiting herd size to carrying capacity and the potential for livestock improvement through selection and breeding. Major investments were made, including fencing the ranch's perimeter and four paddocks, drilling three boreholes, installing a dip and employing a ranch manager. Ten Maasai households (90 people), owning 1,400 head of cattle, were chosen to settle on the ranch, with the agreement that they would dip their cattle weekly, vaccinate their herds regularly, follow a plan of rotational grazing, and restrict their herd numbers to a prescribed limit. Ayuko relates the ranch's subsequent deterioration:

> Wire fences were rendered unserviceable by game animals, and by 1955 the wire and posts had been removed. The families refused to honour some of the commitments, and particularly the one in respect of stock numbers, so that the cattle population increased from the original 1,400 to 2,300 head by 1954; attempts to impose reductions were resisted and finally

> led to four families leaving the ranch with 666 head of their cattle. The remaining six families, having agreed at first not to exceed a maximum of 1,700 head, had increased their herds to 2,441 by 1958. Finally came the drought of 1961 which left the ground bare, forcing all the families to leave. [1980, p. 179]

Though grazing interventions in Maasailand were not successful, in other range areas of Kenya where the grazing schemes existed over a period of years, a regeneration of grasses and an increase in carrying capacities resulted [Ruthenberg 1966]. However, by the early 1960s most of the schemes throughout Kenya were abandoned as disciplined management dissipated with the approach of independence and grazing systems were disrupted by the disasterous drought of 1960-1961. The area under any degree of grazing control fell from 3.7 million ha in 1960 to about 320,000 ha in 1963.

Included in grazing scheme development as well as undertaken as separate projects, water resource improvement became a major area of intervention. Peberdy [1969] notes that between 1945 and 1962, ALDEV was involved in constructing 1,020 seasonal dams, 332 permanent dams, 308 sub-surface dams, 38 rock catchments, 40 masonry weirs, and 72 piping schemes; protecting 54 springs; and drilling 44 boreholes and over 200 wells. But, water development projects frequently only aggravated grazing pressures, since there was not concurrent limitation of livestock numbers to a particular area's grazing capacity. Increases in livestock numbers due to disease control measures had similar mixed effects, prompting opinions that "the provision of drugs which allow cattle in these areas to ravage land which would otherwise be protected by [tsetse] fly is quite unwarrented" [Jones 1959, p. 61].

In addition to grazing schemes, and water and veterinary projects, governmental interventions intended to promote increased sales of livestock from the pastoral areas were instituted. The Kenya Meat Commission (KMC) was established in 1950 and was conferred "exclusive rights for the purchase of cattle and small stock; to acquire, establish and operate abattoirs, meat works, storage concerns and refrigerating facilities; and to slaughter those same stock for export and national consumption of meat, processed meat foods and by-products" [Spink et al. 1980, preface]. KMC was intended to be the principal entity for implementing livestock sector policies, by acting as a price stabilizing agency, buyer of last resort during drought periods, and exclusive exporter of meat for Kenya. Included in its operations was a buying program intended to generate increased sales by pastoralists. But the Commission purchased pastoralists' animals with extreme caution, since it was not authorized to incur financial loss or to undertake unjustifiable risks in the interests of destocking range areas. Unable to pay prices for livestock that private traders were willing to pay, KMC's pastoral operations quickly proved ineffectual.

To more directly address the problem of apparent low livestock marketing levels in pastoral areas, the African Livestock Marketing Organization (ALMO) was formed in 1952. Working in collaboration with the Veterinary Department and authorized to pay subsidized purchase prices if necessary, ALMO assumed KMC's responsibilities for buying African cattle. KMC virtually became a buyer solely of European farmers' and ranchers' slaughter stock, and between 1952 and 1960 sustained a constant annual growth rate of 17 percent, more than trebling its annual volume of throughput from 7,300 mt to 25,500 mt of beef [Kivunja 1978]. Meanwhile, ALMO proved no more successful than had been KMC in competing with private traders in the African areas. Two livestock marketing systems emerged, with private traders purchasing African stock on-the-hoof, and KMC buying the stock of Europeans at controlled prices, paying producers on a quality and dressed-weight basis.

Market dichotomization was reinforced by the lack of well developed marketing channels from pastoral areas to the more populated Highlands, and particularly by movement restrictions necessitated by the risk of disease transmission. More than any other single factor the disease threat to clean areas inhibited the growth of an integrated marketing network. Health of the largely improved dairy herds of the higher-potential areas was the overriding concern, as noted by Jones at the time:

> Since the economy of the country depends mainly on its mixed farming, the safeguarding of the valuable dairy industry from disease inevitably assumes greater importance than the destocking of the pastoral areas. Exotic cattle are much more susceptible to disease than native stock, and this means that disease precautions have to be stringent and that when there is a conflict of interest between the safety of the dairy industry and the free flow of slaughter stock, the interest of the former prevails. [1959, p. 10]

As a consequence of this disease constraint on movement, there existed throughout this period a sizable cattle trade southward from Kenya's Maasailand to markets in Tanganyika (present-day Tanzania excluding Zanzibar).

Thus, as was true for agriculture in general, Kenya's livestock industry was composed of distinctly separate parts, the undeveloped, principally subsistent African production systems and the market oriented operations in the Scheduled Areas (the more neutral title given the White Highlands). In the decade preceding Kenya's independence in 1963, four-fifths of the agricultural sales were by the large farms of the Scheduled Areas, aided by price supports which maintained prices, especially for cereals, well above world levels [Hinga and Heyer 1976]. Though the only livestock products for which there were similar supports were dairy products, by 1960 slaughter stock comprised one-eighth of the Scheduled Area's total agricultural sales [Hazlewood 1979]. Meanwhile, even with the operation of agencies such as ALDEV and ALMO, change was relatively inconsequential in African areas, as exemplified by the fact that from 1946 to 1960 ALDEV resettled only about 11,000 families in its land rehabilitation programs [Odingo 1971]. Pastoral interventions, especially water and disease control programs, tended to increase the quantity but not necessarily the quality of life. Herds multiplied, but market participation was constrained externally by the ever present threat of disease transmission and internally by pastoralists' nonmarket orientation.

Independence came in the wake of the drought and flooding of 1960-1961. During the following years of recovery, livestock production in the pastoral areas received renewed attention, as rural development projects were initiated in various parts of the country [Found 1980; Livingstone 1981]. ALMO was replaced in the newly independent government by a Livestock Marketing Division, and a Range Management Division was also established. Administrative rejuvenation was matched by a more positive outlook regarding the productive potential of the rangelands. Pastoral regions, previously viewed as little more than an economic liability, began to be considered in terms of the positive contribution they could make to bolstering the young nation's economy. To develop and draw forth this contribution became a principal objective of the Kenya Livestock Development Project.

The Kenya Livestock Development Project

The Kenya Livestock Development Project (KLDP), begun in 1968, was intended as a pilot project on which subsequent World Bank livestock projects for African countries would be based [Crotty 1980]. For the Government of Kenya, KLDP is a comprehensive effort to increase rangeland production and bring the long-term goal of increased stratification of the country's livestock industry closer to fulfillment [von Kaufmann 1976]. The activities of KLDP extend beyond pastoral areas, but the formation in these regions of livestock production units through which development inputs could be introduced has been a principal objective. A description of KLDP's overall operations is appropriate, in order to place this objective in perspective.

Briefly, the production units established or upgraded through KLDP are categorized as commercial, company, cooperative, group and individual ranches, and grazing blocks. Commercial ranches, generally in existence before the inception of KLDP, are owned or held on lease by individuals or shareholders numbering from two to several hundred persons. Company ranches, enterprises operated on state land leased from the national government or on trust land administered by district-level county councils, are of three types differing mainly in number of members and degree of public control. The private company ranch is limited to 50 shareholders, who usually live outside the ranch. The public company ranch may have any number of members and is required to make public the subscription of company shares. The directed company ranch is also not limited in its membership, but is required to have an appointed ranch manager, a position frequently filled by a district range officer. Shareholders supply cattle or a cash equivalent on all company ranches. The herd is owned collectively, and profits are shared according to formal agreement.

Cooperative ranches are of two types. The commercial cooperative ranch operates like a company ranch, complete with paid manager and board of directors, and is subject to controls on expenditures and the distribution of profits. All livestock are the cooperative's property. On communal cooperative ranches, individual members are given land at the periphery of the ranch on which to live and keep private herds, in addition to contributing to the jointly owned herd. The allotment of stock units is proportional to a member's shares in the cooperative. Sixty percent of the members are required to reside on the ranch.

Group ranches have been designed for the communally grazed rangelands of the Maasai and Samburu. Since group and individual ranches are the forms of tenurial change found in Maasailand, they are addressed in Chapter IV. Finally, in the arid northeastern part of the country, grazing blocks have been formed. Their development is intended to regulate and control the movements of pastoralists through water investments, controlled grazing, and improved marketing operations [Devres 1979].

These various types of production units are the entities through which the major share of KLDP development loans have been disbursed. Of the funds apportioned under the first phase of the Project (KLDP I), 62 percent was to be for ranch development, 18 percent for equipping the Livestock Marketing Division, 13 percent for surveying and developing of range water resources, and the remaining 7 percent for veterinary, range management, and technical services [von Kaufmann 1976]. By 1974, when the first phase of the Project ended and the second phase (KLDP II) was beginning, 108 ranches had received loans, including 10 company and cooperative, 42 commercial, 41 individual, and 15 group ranches [ILCA 1978].

For the commercial ranches, funds were provided for the development of water resources and stock handling facilities, other capital developments, and the purchase of steers and breeding stock. Similar disbursements were made to the company and cooperative ranches, though often more substantial capital investments were required since many of these ranches were established on previously unoccupied land [Semenye and Chabari 1980]. Water development was given priority under KLDP I on group and individual ranches, and was the target of over half of their allocated funds. Upgrading of herds also was a major area of investment, with loans for improved bulls accounting for 16 percent of the loans to group ranches and 10 percent of the loans to individual ranches [Hampson 1975].

KLDP II, in size and comprehensiveness, was a much more ambitious plan than had been KLDP I. The objectives of KLDP II included the improvement of 2.8 million ha of grazing lands in Northeastern Province (the grazing blocks), and the establishment or improvement of 60 group ranches, 100 commercial ranches, 21 company and cooperative ranches, and 3 feedlots [White 1978]. It was recognized that if KLDP II were to meet production targets, the handling capacity of the Livestock Marketing Division (LMD) would need to be expanded. Investments proposed were expected to increase LMD's capacity for transporting cattle from the northern rangelands to 150,000 head per year, as well as permit broader LMD involvement in Maasailand. As a comprehensive project, KLDP II even contained provisions for the preservation of wildlife reserves. Given the extensive investment planned, development expectations at the time ran high:

> National beef production would be raised by 23,000 tons, a 50 percent increase over present marketed levels. It would also create an estimated 5,000 new jobs, 50,000 people would be engaged on ranches and 90,000 pastoralists would benefit from the general development of the rangeland. [FAO/IBRD 1977, p. 119]

During the first years of KLDP II, it was evident that the development process would be neither as extensive nor rapid as had been projected. Estimates of the number of ranches to be funded were revised downward, mid-Project, to 104 ranches, including 46 commercial, 29 company and cooperative, and 29 group. Altogether, the number of ranches to be funded by KLDP I and II was expected to reach 212, out of an estimated total of 450 to 500 beef ranches in the country, with 250,000 to 300,000 head of cattle affected, out of an estimated range population of 4 million head [ILCA 1978].

By the end of 1978, even the lowered expectations had not been met. Only 32 ranch loans had been fully or partially disbursed under KLDP II, including 15 to commercial ranches and 13 to company ranches [Semenye and Chabari 1980]. Development capital was disbursed at a rate equivalent to about Ksh 940,000 per quarter, half for the purchase of breeding stock, a third for developing water facilities, and the remainder for other improvements [ILCA 1978].

Difficulties, financial and otherwise, impeded progress. Exemplifying the first area of obstruction, between KLDP II's appraisal in 1972 and review reports in 1976, the costs of development inputs had risen dramatically: "Piping costs almost four times, boreholes more than twice, tanks and troughs, firebreaks, fencing and bushclearing, all doubled in cost and dips rose by 50 percent" [ILCA 1978, p. 24]. While rising costs caused delays, the more serious obstacles in pastoral areas have been unforeseen or misinterpreted ecological and social conditions. One example regarding the grazing blocks is related by Bille [1980] and Helland [1980d]. Within each block there was to be a rotational grazing system based on the planned distribution of a large number of constructed storage ponds. In actuality, these ponds did not prevent the continued concentration of cattle, and the alleviation of environmental degradation for which the storage ponds were designed was not attained. As concluded by Helland:

> By modification of the environment by the provision of water, the grazing block project has replaced harsh, direct and efficient natural control mechanisms with man-made, "soft-approach" controls. The dangers inherent in tinkering with the water/pasture/animal balance seem to have been realized but not followed to the logical conclusion of providing the modified system with controls functionally equivalent to the natural ones. [1980d, p. 34]

Summary

The picture that emerges of historic and current efforts to develop Kenya's rangelands is one marked by shortcomings. During the colonial period the development of pastoral areas was hindered by a biased economic commitment in favor of settlers' interests. Rangeland development policies for African areas were eventually initiated, but only in response to increasing ecological deterioration. As the disappointing record of KLDP demonstrates, even recent, well-financed programs have not met with the success expected. Socioeconomic factors, in particular, have hindered advancement. In the following section, the salient problems of interventionism are addressed in general terms, and reasons for the lack of success are posed.

PASTORAL TRANSITION

The history of pastoral interventionism in Kenya described is representative of events in much of Subsaharan Africa. An initial benign neglect of pastoral societies by colonial powers generally gave way to engagement during the second third of the century, predicated on rangeland degradation attributed to pastoral mismanagement. Reliance, on the one hand, upon methods of coercion in which destocking was viewed as an end in itself, and on the other hand, upon purely technological solutions, only invited failure [Baker 1976; Commoner 1972; Helland 1980a].

The understanding of pastoral land use problems matured during the 1960s and 1970s, with recognition of the misuse of resources as but a symptom of the growing maladjustment of pastoral livestock systems due to increasing population densities and limited resources [Baker 1975; Jahnke 1982]. Development of a more enlightened understanding of pastoral issues was aided by the coordination of resource and people based pastoral research efforts, approaches which have been distinct and even antagonistic in their methods and objectives [Aronson 1981; Goldschmidt 1981b]. More recently, coordination of research has found a theoretical as well as practical basis in the farming systems methodology. Examination of pastoralism as a system has required that attention be given to the dynamic interplay among the human, livestock, and environmental components. Only by analysis of interactions within and among these principal components can insights be realized [Anderson and Trail 1981; Dyson-Hudson 1980d; Norman 1982; Rogers 1983; Slovic, Kunreuther, and White 1974; Spedding 1975].

Perceptions of pastoral resource use problems have been enhanced by a systems perspective, but solutions remain elusive. In Kenya, officials admit that today the overgrazing problem is as serious as ever: "Despite warning sounded as far back as 1929 on the overstocking condition in pastoral areas, no satisfactory solution has been found to this problem yet. This has been the greatest failure in Kenya's livestock development history" [Kenya, Republic of 1980a, p. 21].

Unexpected, deleterious outcomes of pastoral interventions have frequently aggravated the problem

[Dyson-Hudson 1980c; Hampson 1975; Helland 1980c; Helland 1980e]. For the Maasai, in particular, Jacobs assesses the "progressive over-peopling, overstocking and overgrazing [as having been] either caused or exacerbated by poorly designed and highly erratic development schemes and policies. . ." [1975, p. 419]. More broadly, in a general review of pastoral development efforts Goldschmidt concludes that the record "is one of almost unrelieved failure. Nothing seems to work, few pastoral people's lives have improved, there is no evidence of increased production of meat and milk, the land continues to deteriorate, and millions of dollars have been spent" [1981b, p. 116]. Meanwhile, communal grazing and livestock mismanagement continue to be identified as the principal problem areas [Dasmann, Milton, and Freeman 1973; Jarvis 1980; Malechek 1982; Norris 1982], and advice to governments that pastoral practices must be modified has become numbingly repetitive [Chemonics International Consulting Division 1977; Doran, Low, and Kemp 1979].

The history of general failure of pastoral interventions is at least partly the result of a larger problem of contradictory objectives and lack of political will at the national level [Chenery 1961; Schaefer-Kehnert and Brown 1973]. For example, the Kenyan Government recognizes that it is in the national interest to bring pastoral lands into market oriented production [Davis 1970; Payne 1976], and yet has hindered this transition by succumbing to the short-term political advantages gained by "cheap meat" and other urban-biased policies [Bates 1981b; Hjort 1981; House and Killick 1981]. The government, as described in Chapter III, is now apparently realizing the shortsightedness of imposing controls which amount to production disincentives.

Conflicts regarding the goals of development programs are less easily reconciled. The generic objectives of development projects in pastoral Africa are succinctly defined by Pratt as "to assist national efforts that aim to change production and marketing systems in pastoral Africa so as to increase the sustained output and yield of livestock products and improve the quality of life of the people of this region" [1980, p. 109]. These objectives are not contradictory, and in fact in the long run each implies the other. But, in the near term, pastoral and national interests may well be in conflict, requiring that priorities be set [Konczacki 1978; Ruthenberg 1980a]. This conflict is evident in criticisms of, for instance, the use of numbers of marketed animals as a criterion in the assessment of pastoral projects [Dahl and Hjort 1976; Galaty 1981a].

Integration of pastoralists into the larger economy is gaining momentum, with or without governmental assistance. Today, no pastoral people in East Africa subsist exclusively on livestock products, and the trend is rapidly toward increasing levels of market involvement, economic diversification and sedantarization [Brandstrom, Hultin, and Lindstrom 1979; Frantz 1975a; Frantz 1975b; Hogg 1980; Johnson 1977; Little 1982a]. Viability is no longer simply a function of the household economy. On top of carrying capacities and livestock productivities, pastoralists' marketing practices and terms of trade with the larger economy are redefining critical population densities [Helland 1980a].

Konczacki states that increased intervention is inevitable if pastoralism "is to survive as a way of life and a method of production" [1978, p. 35]. In a similar vein, others have suggested that pastoralists need continued assurance of their ability to sustain themselves by the pastoral mode of production [Institute for Development Anthropology 1980; Swift 1982]. But this form of assistance, what Galaty [1981b] aptly refers to as the kindly bear-hug of governmental intervention, only prolongs the transition without facilitating the adjustment process. Even with increased intervention pastoralism cannot be preserved. Current and projected population pressures dictate that the only future for pastoral livestock systems is in their transition to market oriented systems [Brown 1963; Ruthenberg 1980b]. The role of interventions must be to encourage and assist the transition to commercialized production by helping pastoralists accommodate the additional risks and uncertainties involved in shifting from a nonmarket to a market oriented set of resource use patterns [Njoka 1979]. It cannot be forgotten that in addition to reconciling forage demand and forage supply, an increasingly difficult task as population pressures mount, the Government of Kenya's concurrent goal in pastoral areas such as Maasailand is higher levels of marketed production.

Modernizing agriculture involves more than modification at the margins [Dorner 1972; Mosher 1969]. Given the fundamental differences between the pastoral mode of production and commercial stockraising, there is truth in statements describing the transition of pastoral livestock systems as a radical revolution and one for which there are no half measures [Cruz de Carvalho 1974; Peberdy 1969; Schneider 1981b]. Even so, socioeconomic change is rarely discrete or absolute. Rather, a process of adaptation and response is taking place in Maasailand and other pastoral regions, and will continue to occur, as both the benefits of market production and the costs of pastoral production rise [Dyson-Hudson and Dyson-Hudson 1982; Salzman 1980]. The critical issue is the accelerating or restraining effects governmental policies and activities have upon the adaptation-and-response process.

CONCLUSIONS

"Resource" is a highly relative concept. As noted by Ciriacy-Wantrup [1963], it changes with the means-end scheme, that is, with the planning agent, that person's or organization's objectives, the state of technology, and existing social institutions. This chapter has delineated the need and potential for an economic transition in Subsaharan Africa's pastoral lands, Kenya's Maasailand in particular, as the resource base is brought into a new means-end scheme. Under the pastoral mode of production, these lands once provided adequately, if capriciously, for the pastoral household's viability as long as its regenerative limits were not exceeded. But population pressures in combination with communal access to resources have made violation of the limits inevitable. Even where stock rates are still maintained at sustainable levels, pastoralism cannot generate the increasing levels of production required by the larger society.

The transition to a means-end scheme in which the planning agent can exert a measure of control over the resource base, can take advantage of productive technologies beyond those which characterize pastoralism, and has market oriented objectives, is occurring in Maasailand. In the past, the misperception of problems and poorly defined objectives have hindered governmental activities designed to assist pastoral development. Today, it is clear that interventions need to be designed and implemented expressly to assist pastoral societies in the transition to commercial production. In the following chapter, ways in which the Kenyan Government is currently affecting the transition process are examined.

3 Governmental Intervention and the Pastoral Transition

The transition for the pastoral Maasai to increasingly market oriented livestock production cannot be considered apart from the government's commitment to directed change [Jenny 1980]. With respect to livestock, this commitment translates into a goal of increased production levels. As explicitly stated in the National Livestock Development Policy, a document self-acclaimed as "the most definitive policy on livestock development since independence":

> The most basic objective will be to increase productive investment and growth in output of livestock products. In the short term (1980-1983), the objective will be to avert the projected shortages in meat and milk. In the longer run, the objective will be to increase animal products to feed an increasing population. [Kenya, Republic of 1980a, p. 12]

The development of a stratified livestock industry, that is, cow-calf operations in the drier rangeland areas concentrating on the production of immatures with more humid regions used for the finishing of slaughter stock, envelops the government's various strategies for accomplishing the long-term objective. Stratification necessarily implies the integration of pastoral peoples such as the Maasai into the production chain, that is, promotion of the transition process discussed in Chapter II. In this chapter some of the principal forms of governmental intervention which are influencing the process for the Maasai are examined.

In the first section, pricing and marketing interventions, operations of the Livestock Marketing Division (LMD), and structural changes in Kenya's livestock industry are addressed. Discussion of three additional areas of intervention comprises the second section: disease control, upgrading of livestock, and wildlife management. Policy recommendations for expanding production require recognition of the interrelationships and relative significance of these various existing forms of intervention.

PRICING AND MARKETING POLICIES AND A CHANGING LIVESTOCK INDUSTRY

Price Controls

There are basically four points of exchange in a livestock industry, between producer and trader, trader and wholesaler, wholesaler and retailer, and retailer and consumer. In Kenya, the impact of price controls is felt at each of these points.

Structure of controls and price levels. Price discovery is usually by auction or through one-to-one negotiation between buyer and seller at local-level livestock markets in Kenya. Prices agreed upon in such transactions are not directly administered. However, they are indirectly influenced by controls at subsequent links in the marketing chain, as well as by prices paid to producers by the Kenya Meat Commission (KMC). The latter set of prices, fixed by the Minister of Agriculture, can be considered national floor prices.

During the 1970s, real prices per kg paid producers by KMC remained essentially unchanged for cattle, while for small stock, especially goats, they declined dramatically; mean cattle carcass weights fell by nearly 20 percent, while small stock mean weights increased by about the same proportion (Table 3.1). Average cattle carcass weights in Kenya declined between 1977 and 1980, contrary to worldwide trends (Table 3.2). African and developing nations in general maintained constant carcass weight levels during this period and developed nations recorded increasing average carcass weights. The decline in Kenya's slaughter weights can be attributed to internal factors, particularly meat price controls.

At subsequent points of exchange, wholesale and retail prices for beef, lamb, mutton, and goat meat are fixed by the Minister of Finance, as empowered by the Price Control Act. Price Control (Meat) Orders, the most recent of which was gazetted in February, 1983, are based on KMC producer prices plus estimated costs of processing at KMC [ILCA 1978]. The Order is comprised of Five Schedules. The First and Second Schedules fix the maximum wholesale prices payable to KMC and to "all other abattoirs," respectively. The prices per kg set for beef are by grade, on a forequarter and hindquarter basis. Lamb, mutton, and goat meat are not graded. Table 3.3, which shows wholesale beef prices in Kenya in relation to those in other parts of the world, 1978 to 1981, indicates that

Table 3.1. Percentage changes between 1972 and 1980 in real prices and dressed weights of livestock purchases by the Kenya Meat Commission

Livestock	Change between 1972 and 1980: Mean Real Price per kg Dressed Carcass[a]	Change between 1972 and 1980: Mean Dressed Carcass Weight
	percent	percent
Cattle	+ 1.2	-19.6
Calves[b]	--	--
Lamb	-18.0	-17.6
Sheep	-22.4	+23.1
Goats	-42.4	+20.0

Source: Based on Kenya Meat Commission data.

[a]Deflated by the general consumer price index for Nairobi [ILO 1982].

[b]No recorded purchase of calves in 1980.

Table 3.2. Average cattle carcass weight in Kenya in comparison with world averages, 1977-1980

	Average Weight			
Country/Region	1977	1978	1979	1980
	---------kg per carcass---------			
Kenya	140	140	130	125
Africa	138	140	140	140
All developing countries	161	161	161	161
United States	246	255	269	272
All developed countries	213	216	220	219
World	195	196	198	198

Source: FAO [1980, 1982a].

Table 3.3. Kenyan wholesale beef prices in relation to world price levels, 1978-1979 and 1980-1981

	Wholesale Beef Prices			
Country/Region	1978-1979		1980-1981	
	-dollars per kg-	-index[a]-	-dollars per kg-	-index-
Kenya[b]	1.19[c]	0.86	1.38[d]	1.00
United States[e]	1.32	0.96	1.44	1.04
EEC[f]	1.55	1.12	1.72	1.25
Australia[g]	1.23	0.89	1.56	1.13
Argentina[h]	1.13	0.82	1.59	1.15

Source: Calculated from Table 3.2, and FAO [1982a].

[a]Kenya's wholesale beef price, 1980-1981 = 1.0.

[b]Based on FAQ grade prices; prices for KMC and All Other Abattoirs weighted equally.

[c]Exchange rate: Ksh 7.60 = $1.00.

[d]Exchange rate: Ksh 8.23 = $1.00.

[e]Steers, 900-1,100 lb liveweight.

[f]Adult, weighted average liveweight.

[g]Brisbane, oxen, 301-320 kg slaughter weight.

[h]FOB; all beef, export average unit value.

Kenya's prices have been on the low side throughout this period.

The Third, Fourth and Fifth Schedules fix maximum retail prices, without reference to grade. Five commodities are specified in the Third Schedule: beef with bone; beef without bone; mutton/goat meat; liver, heart, tongue and kidney; and tripe (matumbo). Prices set for each of the five commodities vary by locality. Schedules Four and Five set prices for special cuts of meat in Nairobi and Mombasa, and in "other areas," respectively. Clearly, meat prices in Kenya are closely controlled at the wholesale and retail levels. The impact administered prices have had on production is equally apparent.

Effects on livestock production. Meat price controls have severely handicapped livestock development in Kenya. Several studies undertaken during the late 1970s detailed major discrepancies between official price levels and actual production-marketing costs. For example, McArthur and Smith [1979] report in one study, based on livestock prices collected at 16 markets during 1978 and 1979, that the estimated market price (saleyard price plus processing costs and profit, minus the value of the organs and nonmeat products) was in all cases above the maximum regulated retail price. They also found that the range in estimated liveweights of 172 head of cattle at the Kibiko Holding Ground, site of a major livestock market serving Nairobi, was 105 to 350 kg. Liveweight prices averaged Ksh 5 per kg, and ranged from Ksh 2.86 to Ksh 6.39. As they relate:

> Assuming a 42% dressing percentage, this would imply Ksh 11.20/kg CDW and with allowance for cost and profit, and the value of the fifth quarter, this would produce a retail price (assuming no additional wholesale margins) of Ksh 13.10/kg, compared to the regulated retail price of meat bone-in of Ksh 10.30/kg in Nairobi. That this price structure does prevail is confirmed by interviews at Dagoretti market in November, 1978, when hindquarters were shown to be selling at Ksh 12.00/kg wholesale. [McArthur and Smith 1979, p. 18]

In addition to official livestock and meat prices being held below their equilibrium levels, a number of pricing inconsistencies have contributed to a general decline in the quality of meat marketed. As described above, while wholesale prices are fixed according to grade, retail prices are not. Butchers have higher net returns per kg when it is the lower-grade carcasses which are sold in the form of special cuts [Matthes 1979]. Thus, they are encouraged to purchase low and medium grade carcasses, a situation that has enhanced the demand for lower-grade rather than higher-grade animals.

A second area of pricing inconsistency arises due to the range of meat products included in the category "beef with bone." Not only does bone content vary among cuts from the same carcass, but bone proportions across grades can differ considerably. Bone as a percentage of dressed weight can be as much as 45 percent greater for a commercially graded animal as for one of choice grade [Spink et al. 1980]. Since the "beef with bone" retail price is the same for one kg from any carcass, the market for lower-grade (cheaper) animals has expanded, while that of higher-grade animals has contracted, helping to explain the decline in cattle carcass weights (Tables 3.1 and 3.2). These observations illustrate why price controls, compounded by grading inconsistencies, have been counterproductive to increased production and, in particular, to the upgrading of the national herd. Moreover, as is true wherever prices are administered, they have invited evasion. Butchers who market higher-quality grades of meat often illegally ignore controls and charge prices above those authorized, or avoid price controls by tactics such as selling meat directly to restaurants [Devres 1979; ILCA 1978].

McArthur and Smith [1979] suggest that governmental intervention in the pricing and marketing of meat might be said to be of benefit to the economy if it leads to the following:

(i) improvement in resource allocation,

(ii) reduction in price, volume, and income instability,

(iii) increased economic self-sufficiency in meat production, or

(iv) provision of low priced meat to low income consumers, provided that this does not inhibit supplies or lead to resource misallocation.

In considering whether Kenya's meat price controls have helped to attain any of these four conditions, it is noted that:

(i) Price controls have encouraged the misallocation of resources. A favorable pricing structure is generally considered vital for ranch development but, in Kenya, ranching investments have not been forthcoming [Cronin 1978; de Wilde 1980b]. Production has stagnated in the face of rising demand, suggesting that productive resources have not been utilized in their socially defined best use.

(ii) Retail prices for beef lagged behind the general level of prices during the 1970s. Between December, 1971, and July, 1979, the upper income food price index increased by 145 percent and the middle income index rose by 121 percent. During the same period, the controlled price of beef with bone rose by only 77.7 percent. Put another way, the relative price of beef, compared with a basket of other food items, fell by some 24 percent during the 1970s [UNDP/FAO 1979]. A decline in the relative price of meat might be welcomed, until one considers the impact on producers' incomes. Declining price trends are not a mark of stability.

(iii) Projections presented in Chapter I indicate the distinct possibility of a deficit in meat supplies by 1990. A country with a promising meat export potential in the early 1970s, Kenya today faces impending shortages.

(iv) Lower-income urban consumers, when compared with their rural counterparts, comprise a small fraction of Kenya's total population. At the extreme, the ratio of rural/urban poverty incidence is on the order of 10 to 1 [FAO 1982b]. The urban poor are certainly interested in low meat prices. However, the much more numerous rural poor are also lower-income producers, and as such do not benefit in the long run from cheap meat policies. Essentially, the income of urban consumers is subsidized at the expense of rural producers. Price controls constitute an income transfer from herdsmen and farmers to urban dwellers.

Notwithstanding the ramifications of altered price ratios between meat and other goods and services, a complete freeing of meat prices would be the most significant action that the government could take to increase production, particularly of higher-grade meat. This conclusion has been reached repeatedly in livestock studies in Subsaharan Africa [Simpson 1976; Sullivan 1984; UNDP/FAO 1979], and the logic behind it is readily depicted graphically. Figure 3.1 shows a situation in which price of a commodity (type of meat) is held below its equilibrium level, that is, the price at which quantity supplied would equal quantity demanded (P_e). At the controlled price level (P_c), quantity supplied (OQ_S) is less than quantity demanded (OQ_D). In other words, production is depressed by insufficient price incentives and demand is inflated by "cheap meat" prices, with upward pressure on prices manifested through illegal attempts to evade price controls. Demand is ever excessive, since only the quantity OQ_S is available. In the longer term, producers are not only discouraged from making investments to expand production, but resources are shifted to enterprises or end uses yielding higher perceived returns. For a pastoralist household, this effect essentially results in less production for market and reinforcement of the traditional, subsistence orientation.

It is ironic that price contols are touted by the government as contributing to the national goal of poverty alleviation, since in the long run they clearly restrain growth and development. Realistically, however, price distortions cannot be easily eliminated, for they are a reflection of the distribution of economic power and not an incidental development arising from mistaken policies

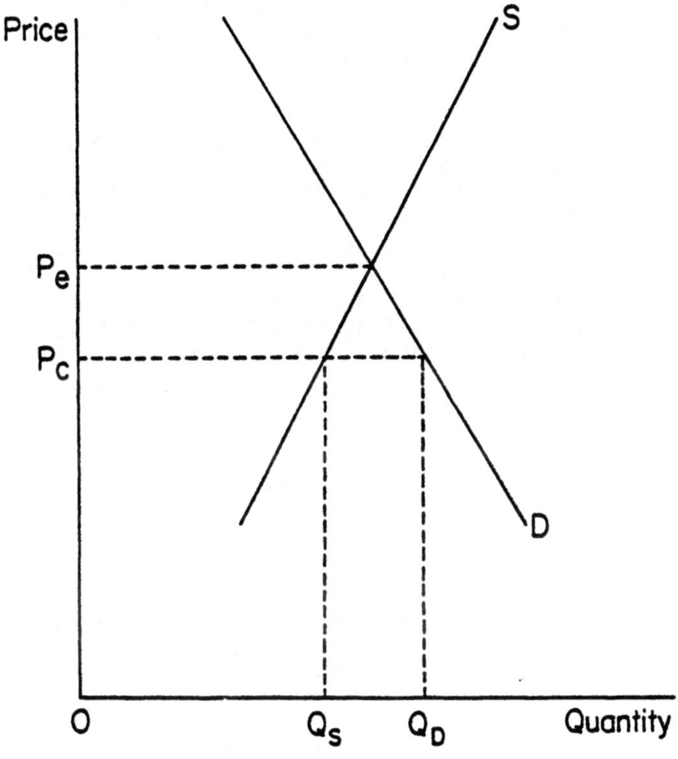

Figure 3.1. The effect of price controls on quantity supplied and quantity demanded of a commodity

[Stewart 1981]. Significantly, the government has recently begun to assume a policy position more responsive to producers' needs, a policy shift also occurring elsewhere in Subsaharan Africa [Bates 1981a; Kenya, Republic of 1980b]. For example, it was recognized in the Development Plan, 1979-1983, that for the agricultural sector there were "limited possibilities for large increases in producer prices except for livestock" [Sharpley 1981, p.318, emphasis added]. Meat prices have been increased three times in the past four years and, as indicated, the grading of meat has been simplified. Whether these increases will influence production positively largely depends upon concurrent changes in other prices. At the least, price increases of the early 1980s suggest that there has been a modification of the power base of policy makers, and the deleterious effects which low prices have had on production can no longer be disregarded politically.

The establishment of a price environment favorable to the transition of Maasailand will hinge on the political fortitude of the government's commitment to provide producers with needed incentives. At the least, given Kenya's high income elasticity of demand for beef described in Chapter I, the relative prices of higher-quality cuts should be increased. As it is, prices will continue to gradually rise despite controls, with whatever presumed positive effects in restraining inflationary trends and making meat more available to Kenyans more than offset by the negative impact on production [Killick 1979]. The importance of prices to expanded livestock production is further explored, along with other factors, in the following examination of the government's livestock marketing interventions.

Operations of the Livestock Marketing Division

Marketing conditions are central to Maasailand's livestock development [UNDP/FAO 1979]. The government's principal involvement in livestock marketing in pastoral areas is through the operations of the Livestock Marketing Division (LMD). The Division's activities in Maasailand have been relatively limited, but it is instructive to examine LMD's extensive marketing role in northern Kenya, in its impact upon Maasai producers and on the transition process.

LMD's responsibility is foremost the prevention of disease transmission in the movement of livestock. To facilitate the flow of cattle from Kenya's northern rangelands, stock are purchased by LMD and transported to holding grounds, where they are either treated for immediate transport to slaughterhouses or quarantined for five to six months before delivery for further growing to ranches located in disease-free parts of the country

[Kenya, Republic of 1980a]. Contagious bovine pleuro-pneumonia (CBPP), a bacterial illness enzootic in northern Kenya and transmitted by direct contact between infected and healthy animals, is the main disease that has necessitated these quarantining procedures.

The quarantine is a lengthy process due to the series of tests which must be conducted. The financial costs and risks involved in the quarantining process are considered too great for private traders to bear and, therefore, the government provides the needed service [Livingstone 1975]. Cattle are tested immediately upon arrival at one of the holding grounds. If clean, they are retested twice at intervals of six weeks. If there is still no sign of disease, they are released for movement. However, if there are any reactors in the herd, those animals are removed and slaughtered, and the process must begin once again for the remaining animals [White 1978].

LMD's buying program in the northern rangelands during the 1960s was overwhelmingly determined by the demand for feeder stock by ranches in higher-potential disease-free areas of Kenya. KMC, with which LMD closely coordinated its purchasing program, relied predominantly at that time on supplies of higher-grade cattle from mixed farms and large-scale ranches. Quotas were issued for slaughter stock from pastoral areas only when there were shortfalls in supply [White and Meadows 1980a]. However, as KMC's throughput declined in the late 1960s, it increased the number of slaughter stock purchased from pastoral areas, and LMD was the principal supplier.

During the early 1970s LMD's total annual purchases were averaging 50,000 head, and over the period 1975 to 1980 were projected to rise to over 100,000 per year, on the assumption that there was a large surplus of young male cattle in the northern rangelands. By way of comparison, Kenya's subsistence and commercial rangeland offtake during 1975 was estimated to total 201,000 head, about 15 percent of national offtake [White 1978]. In actuality, LMD's annual purchases from the northern rangelands proved highly variable from year to year, never reaching the projected levels. The steadily expanding supply of feeder stock which had been envisioned did not occur. Thus, LMD has aptly provided the necessary quarantining service, but has been relatively ineffective in promoting sales.

Inevitably, disease control regulations related to marketing of livestock are costly and controversial [Heyer 1976b]. Throughout the 1970s, LMD operated at a substantial deficit, which rose from Ksh 60.80 per animal in 1972/73 to about Ksh 530 in 1978/79 [Kenya, Republic of 1980a], with the CBPP quarantining operations the major expense. Besides direct quarantining costs, there is the risk of physical loss, as illustrated by events at the Isiolo Holding Ground in 1973/74. Very poor rains at a time when 20,000 head of cattle were undergoing quarantining from the previous year led rapidly to overgrazing, and many of the cattle died despite supplementary feeding efforts.

Also contributing to its financial difficulties, LMD is obliged to act as the government's buyer of last resort in rangelands during times of drought. The consequences of this obligation are evident in Table 3.4, which shows sales of feeder and slaughter stock for three consecutive years. During 1976/77, at the end of a time of drought, the number of head sold by LMD was about six times the numbers sold during the following two years. Slaughter stock greatly outnumbered feeder stock during the drought year, whereas during the subsequent years the proportion of feeder to slaughter stock was more in line with LMD's supply expectations of a 7 to 3 ratio [Meadows and White 1980a]. During the early 1970s, LMD was intended to operate at a profit, but at present it is more realistically expected only to cover its operating costs. The financial health of LMD is significant to Maasailand's livestock development, since LMD cattle from the north are increasingly purchased by Maasai producers for further grazing. The importance of northern feeder cattle to Maasai production is exemplified by the operations of the sampled producers, described in Chapter V.

Several studies conducted at the beginning of the 1970s indicated inefficiencies in LMD's purchasing and transporting procedures [Kenya, Ministry of Agriculture 1970; Kenya, Ministry of Agriculture 1971]. More recently, Njiru's account of difficulties encountered by the Rendille, a pastoral people of the northern rangelands, in their sale of cattle to LMD, illustrates the types of problems that have persisted:

> Although LMD purchases animals at Marsabit the auctions are not well publicized and LMD policy changes so often that very few traders rely on selling to LMD. There are several reasons for this: (a) they only give about two weeks notice, and (b) the type of animal bought is subject to so many policy changes that it is difficult [for the private traders] to buy animals for sale to LMD because what they want changes too often . . . . .impromptu changes in animals wanted is damaging LMD's image as a reliable outlet and it is no wonder that traders still trek their animals to Isiolo and other towns for prices which are not significantly better than those of LMD and at greater risk. [1982, p. 30]

Table 3.4. Sales of immature and slaughter stock by the Livestock Marketing Division, 1976/77 to 1978/79

Age of Livestock	Quantity and Proportion of Year Total					
	1976/77		1977/78		1978/79	
	-head-	-percent-	-head-	-percent-	-head-	-percent-
Immature	11,697	27	5,078	74	6,096	88
Slaughter	31,997	73	1,752	26	859	12
Total	43,694	100	6,830	100	6,955	100

Source: UNDP/FAO [1979], Table 11.

Such inefficiencies indirectly affect, through undue costs and delays in the supply of feeder stock, the transition in Maasailand to increased market oriented production. On the other hand, LMD has not had a major direct impact on Maasai producers. Since Maasailand is officially free of CBPP and does not require quarantining services, the Division's presence has been mainly felt in the development of holding grounds and buying centers, activities which it has proposed to expand [Devres 1979]. Direct participation in the marketing process has been suggested by White [1978] and others, under the assumption that government intervention is needed to supplement private sector activities if offtake is to be increased. This assumption is critically examined in Chapter VII.

LMD's operations have lent momentum to the process of stratification of Kenya's livestock industry, even though quarantining requirements have resulted in a flow far from smooth. The stratification which is taking place between pastoral zones, with Maasai producers using their lands to "finish" cattle purchased from the north, is indicative of their increased market involvement. In the following sub-section, two trends are described which underscore the responsiveness of Maasai and other livestock producers to changing marketing and pricing circumstances and the impact of these responses upon the livestock industry.

Structural Change

Decline of the Kenya Meat Commission. The growth and expansion of the Kenya Meat Commission (KMC) described in Chapter II was not only unsustained after Kenya's independence; it, in fact, experienced a precipitous decline. The Commission had recorded a loss of Ksh 700 million by 1967, after showing a profit of Ksh 60 million in 1961 [Kivunja 1978]. Deficits became the norm throughout the 1970s, and KMC's solvency depended upon the regular infusion of government funds. Livestock purchases by KMC were highly variable throughout the period 1967-1980, dropping off substantially in the later years, such that by late 1980 it was operating at only two-fifths of its weekly production capacity of 750 mt. The decrease in all forms of meat disposal between 1977 and 1980, and the fall-off in sales of hides, skins, and by-products during the same period indicate the severity of the decline (Tables 3.5 and 3.6). For example, the quantity of meat sold in 1980 was only about one-fourth of 1977 sale levels.

KMC's problem of securing sufficient throughput for profitable operation can be largely attributed to increased competition from private abattoirs. In the early 1970s KMC's legal monopoly over commercial slaughter was relaxed, and private slaughterhouses were officially permitted to operate for the first time. They were established for the most part on the outskirts of Nairobi

Table 3.5. Form and quantity of meat sold by the Kenya Meat Commission, 1977-1980

Form of Sale, Type of Meat	Quantity			
	1977	1978	1979	1980
	--------1000 kg dressed weight-------			
In carcass form				
Beef, local	7,950	3,153	2,314	2,451
export	1,844	618	428	102
Veal	53	39	37	3
Mutton	139	149	182	80
Lamb	58	6	1	9
Goat meat	39	75	22	14
Canned beef	7,131	2,487	2,461	2,288
Cut beef	4,096	741	643	899
Total	21,310	7,268	6,088	5,846

Source: Kenya Meat Commission.

Table 3.6. Quantity and value of hides and skins and by-products sold by the Kenya Meat Commission, 1977-1980

Item		1977	1978	1979	1980
Hides and skins					
Quantity,[a]	local	61,554	44,006	34,373	46,701
	export	130,534	36,288	48,810	3,107
	Total	192,088	80,294	83,183	49,808
Value,[b]	local	226,928	107,472	85,565	135,206
	export	413,652	147,238	294,009	7,388
	Total	640,580	254,710	379,574	142,594
By-products					
Quantity[c]		5,223	1,744	1,836	1,350
Value[b]		616,652	209,972	239,669	250,262

Source: Kenya Meat Commission.

[a]Numbers.

[b]Kenyan Pounds (One Kenyan Pound = Ksh 20).

[c]Metric tons.

and to a lesser degree near other urban centers, in support of Mittendorf's [1978] observation that the location of abattoirs is dependent upon the comparative costs of transporting livestock and meat. With Kenya's low meat and high fuel prices, the costs of establishing a cold chain operation from a slaughterhouse located in a producing area were and continue to be prohibitively high.

It was intended that KMC concentrate on Kenya's export market, while local urban markets would be supplied by the private firms [Meadows and White 1979]. As a result, the private abattoirs quickly dominated domestic sales and, when exports foundered, KMC's uncompetitiveness became readily apparent. Livestock suppliers were attracted to the private firms by their willingness to pay higher prices than KMC, purchase on a liveweight basis instead of by carcass weight, and offer immediate rather than postponed payment. For Maasailand, the proportion of total commercial offtake sold to KMC declined from 69 percent in 1972, to 27 percent in 1977 [White and Meadows 1978]. By 1977, private abattoirs were supplying an estimated three-fourths of Nairobi's beef, half of which was bought directly from producers by butchers, while the other half was bought by stock traders and wholesaled to butchers. This change in the slaughter market channels has occurred for all types of producers. Even Kenya's large-farm sector, traditionally the principal source of livestock for KMC, is increasingly selling to private slaughterhouses and their agents [Kenya, Ministry of Economic Planning and Development 1980a].

KMC's failure to retain a competitive stance is evident in Table 3.7, which shows estimates of the marketing costs and margins in 1978 for the three main channels supplying beef to Nairobi. At that time, KMC's slaughtering and processing costs amounted to the equivalent of \$ .27 per kg dressed weight, which was estimated to be at least 10 times that of the private slaughterhouses [Kenya, Ministry of Agriculture 1978]. Whereas the income from by-products usually paid for all operating costs at the private slaughterhouses, they covered only 36 percent of operating costs for KMC [ILCA 1978]. Moreover, less stringent meat inspection practices than at KMC resulted in a lower rate of carcass condemnation--and lower associated costs--for the private firms.

Three-fourths of KMC's cattle purchases were bought on a liveweight basis during 1979, in an attempt to increase throughput. This strategy helped to raise production volume but the operation was not profitable due to the increase in purchasing costs. Spink et al. [1980] concluded that a "premium" of Ksh 187 per animal was paid, an amount equivalent to the purchase of an extra 11,360 cattle of similar grade, or nearly four weeks supply at 60 percent throughput (Table 3.8). Other proposals have recommended KMC's reorganization in order to make it more

Table 3.7. Costs and returns for the three main beef supplying channels for Nairobi, 1978

Item	Costs and Returns		
	Producer-KMC-Butcher	Producer-Trader-Butcher	Producer-Cooperative-Butcher
	---------------dollars---------------		
(1) Purchase price[a]	111.09	140.66	140.66
(2) Transport costs, farm gate-slaughter-house[b]	6.91	6.55	6.55
(3) Feeding costs	--[c]	--	--
(4) Losses in weight	not known	not known	not known
(5) Slaughtering and processing costs	33.41	3.84	2.56
(6) Carcass wholesale price, 125 kg	135.87	136.32	136.32
(7) Hides and offals wholesale price	13.75	19.18	21.74
(8) Total wholesale price	149.62	155.50	158.06
Profit margin (8)-(1)-(2)-(3)-(4)-(5)[d]	-1.79	4.45	8.29

Source: Kenya, Ministry of Agriculture [1978], Table 3.

[a]Mean purchase prices for 1978.

[b]Assumed 100 km by rail.

[c]Included in slaughtering and processing costs.

[d]Includes remuneration for own labor and capital.

Table 3.8. Cost of liveweight purchases by the Kenya Meat Commission in comparison with cost of equivalent dressed weight purchases, 1979

Item	Cost	
	Actual Liveweight Purchase	Equivalent Dressed Weight Purchase
	----------------Ksh---------------	
Total cost	32,085,735.	24,808,817.
Mean cost per carcass	826.	639.
Mean cost per kg dressed weight	7.75	5.99

Source: Spink et al. [1980], Table IV.6.1.

competitive with private abattoirs [UNDP/FAO 1979]. However, organizational changes will be ineffective if basic pricing problems are left unaddressed. Unless KMC can pay higher prices for livestock, and do so promptly, private slaughterhouses will continue to increase their market share.

The growth of private slaughterhouses at the expense of the KMC is a phenomenon which has both encouraged and depended upon increased commercial offtake from Maasailand. KMC's long-held purchasing bias in favor of higher-grade, disease-free animals forced pastoralists to seek other outlets for their slaughter sales. Even though KMC has broadened its purchases in order to bolster a declining throughput, Maasai producers in general have not been drawn away from the higher and more immediate returns earned by selling stock to private buyers. The marketing preferences of Maasai producers reflect their responsiveness to price advantages.

Failure of feedlots. During the past two decades, increased stratification of Kenya's livestock industry is thought to have been impeded by the lack of sufficient finishing facilities for absorbing rangeland immatures [Pratt and Gwynne 1977]. As described for the districts of Maasailand in the mid-1970s:

> If Kajiado and Narok were stocked with 1.0 million head (compared to 1.5 million now), in order to stabilize the population, fattening arrangements would have to be made for 180,000 young steers and heifers annually, and this number alone would nearly fill up the present commercially ranched areas. [IBRD 1977, Annex 6, p. 15]

In order to provide the projected finishing capacity necessary to accommodate feeders produced in the range areas and, in particular, a stable base upon which to establish the country's meat export market, the development of feedlots was given high priority by the government in the late 1960s. The Beef Industry Development Project, funded by the UNDP and FAO beginning in 1968, was set up to determine the optimum intensive stock feeding and management techniques. The Project was largely motivated by the expectation that Kenya's maize production would exceed domestic demand, allowing increasing amounts of maize to be exported either directly or, preferably, as livestock products [Schaefer-Kehnert 1971]. According to the Project's initial findings, high quality beef could be produced from zebu stock utilizing combinations of maize silage, ensiled maize grain and molasses. In test trials, 70 days in feedlot conditions raised the edible meat content of a low quality carcass by 50 percent, and did so profitably [FAO/IBRD 1977].

Intensive feeding offered Kenya the possibility of gaining access to the European market, as it would permit quarantine control of cattle within a well-established disease-free area over a minimum period of 90 days. By 1972, world market prospects for beef probably looked better than for all other Kenyan export industries except tourism [Newberry 1976]. Since intensive feeding operations were also viewed as a solution to the overgrazing of range areas, they were optimistically supported as a means of simultaneously expanding rangeland production and the nation's export earnings [Auriol 1974]. As related by Squire [1976], there were early indications that this optimism was well-founded, and by the early 1970s a number of large-scale farmers in the country's disease-free area had constructed feedlots [Newberry 1976]. A study was even conducted to determine the feasibility of establishing a feedlot in a higher-potential region of Narok District [Squire 1975].

The profitability of the feedlots depended on the relative prices of feeders, finished stock, and feed. As Schaefer-Kehnert cautioned at the beginning of the 1970s, "the price differential between feeder and fat steers is extremely favourable in Kenya at present, and the question is whether this will continue" [1971, p. 11]. Unfortunately, price ratios during the 1970s did in fact move distinctly against feedlotting of cattle, as national grain surpluses turned into deficits. By the end of the decade, the Project's designers found that most of the nation's feedlots had been forced to close due to unprofitable beef/feed price ratios [UNDP/FAO 1979].

The uneven performance of the animal and meat export industry during the 1970s is indicated in Table 3.9. Canned meat exports increased from 3,130 mt in 1970 to 5,776 mt in 1976 [Simpson and Mirowsky 1979]. However, the mid-1970s was a period of drought in Kenya, and the increase in exports during these years can be largely attributed to forced sales. The export industry prospered to the degree that it did, not because of a well established and expanding market for feedlot-finished steers, but rather due to a drought-induced culling of the national herd. The value of exported meat and meat preparations as a proportion of Kenya's total exports fell from 4.2 percent in 1972 to 0.7 percent in 1979 [Kenya, Ministry of Economic Planning and Development 1980b].

The role of feedlots in Kenya has yet to be decided. Maize, the mainstay of the Kenyan diet, will not be in sufficient supply in the near future to be used as the principal feed ration. Moreover, the supply of feeder steers generally fluctuates out of phase to demand when feedlots rely upon maize and other conventional fodder rations [Sandford 1983]. But profitable feeds developed from treated agricultural and industrial waste materials

Table 3.9. Livestock and meat exports for Kenya, 1971-1979

Year	Live Animals, Chiefly for Food	Meat and Meat Preparations
	-----head-------	-----1000 kg------
1971	549,473	7,631
1972	711,209	11,024
1973	1,018,051	6,664
1974	1,376,920	6,780
1975	1,355,728	8,280
1976	2,056,755	9,775
1977	758,216	9,344
1978	227,489	3,033
1979	139,964	2,643

Source: Kenya, Ministry of Economic Planning and Development [1980b], Table 53(b).

are used in other countries and may increasingly be utilized in Kenya [Naga and El-Shazly 1983; Preston 1976; Ranjhan 1978; Shah and Muller 1983]. Such rations, for example, have been the basis for reported successes in the fattening of zebu and crossbred cattle in Ethiopia [O'Donovan 1979]. Ultimately, utilization of by-products is a function of opportunity costs and technical conversion factors. Kenyan livestock can be expected to increasingly be fed on crop residues as prices of concentrates and commercial forages escalate [Potts 1983; Said et al. 1983]. In addition to the development of alternative feeds, the price spread for cattle will be of fundamental importance to the emergence of a national feedlot industry. Profitable feedlotting conditions will depend largely upon a favorable price differential between feeder and finished steers [Schaefer-Kehnert 1978].

The recent histories of KMC and the Feedlot Project exemplify the significance of pricing and marketing changes to the structure of Kenya's livestock industry. KMC's uncompetitive prices led to private abattoirs dominating the domestic market (while the export market languished), providing the Maasai with greater incentives to sell their slaughter stock. Unprofitable beef/feed price ratios have stifled a nascent feedlot industry. Most cattle continue to be finished on pasture, a situation unlikely to change as long as low-cost extensive grazing resources are available [Bishop 1975], or frequently are simply sold and slaughtered as unfinished stock. Clearly, changes in price relationships and marketing opportunities like those described here significantly influence the transition to market oriented production by Maasai producers.

Summary

The negative impacts which pricing policies have had on Kenya's livestock production and the transition process in Maasailand have been described. Meat price controls, imposed in order to restrain increases in the cost of living and, in particular, to benefit urban consumers [Bates 1981b; Fenn 1977], epitomize policy formation as very much a "part of the fabric of social struggle" [Stewart 1981, p. 75]. However, arguments that higher food prices especially hurt lower-income (particularly urban) consumers are usually based on the short-term income effects of higher food prices, and ignore relatively short-run adjustment processes such as shifts in demand and increases in wages [Brown 1978]. A fundamental distinction must be made between measures such as price controls that artificially lower prices and measures such as on-farm investment and infrastructural development that lower prices by lowering the real costs of production.

Achievements and shortcomings of the Livestock Marketing Division in attempting to promote the livestock industry's stratification demonstrate the strengths and weaknesses of direct marketing interventions by the government. LMD's focus on increasing the flow of feeder stock from Kenya's northern rangelands has significantly influenced the expansion of feeding enterprises in Maasailand.

Pricing policies and marketing interventions represent major governmental activities intended to channel and assist the evolution of Kenya's livestock industry. Underlying determinants of supply and demand ultimately set the limits and provide the opportunities for development. The pervasive influence of changes in prices and price ratios on industry structure, and specifically on Maasai producers' orientations, has been illustrated by reference to two major occurrences during the 1970s: the decline of the Kenya Meat Commission and concurrent rise of private abattoirs, and the demise of a short-lived national feedlot project originally intended to firmly establish Kenya as a beef exporter. In the following section additional principal areas of governmental intervention are examined.

DISEASE, BREEDING AND WILDLIFE INTERVENTIONS

Controlling Disease

Disease may be defined as any endogenous condition that results in suboptimal growth, production or reproduction, or death [Ellis and Hugh-Jones 1976]. From the devastating rinderpest pandemic of the 1890s, to present-day outbreaks of foot-and-mouth disease, East Coast fever, and other afflictions, diseases have been a major obstacle to increased livestock production in Maasailand. The quarantining activities of LMD demonstrate the priority which the government has given to implementing effective disease control measures for Kenya as a whole. For some diseases such as CBPP, rinderpest, anthrax, blackquarter, and foot-and-mouth disease, vaccination has helped to curtail losses. Various other disease problems are also controllable; for example, emaciation and anaemia caused by helminths and other endoparasites can be kept in check by the regular drenching of stock [McCauley 1983].

Many diseases which affect Maasai livestock, such as trypanosomiasis, are more difficult to subdue. Large areas of bushy rangeland are infested with the tsetse fly which transmits trypanosomes to livestock. Research is proceeding on ways of increasing animals' levels of trypanotolerance, that is, reducing their susceptibility to trypanosomiasis [Murray et al. 1979]. In the meantime, control of this disease has depended mainly on bush control and wildlife restrictions, and currently on the application of insecticides [de Vos 1978; Pratt and Gwynne

1977]. None of these measures can be considered a long-term solution.

Malignant catarrhal fever, spread by wildebeest at the time of calving, is another example of a fatal disease for which an effective prophylactic treatment has yet to be developed. Livestock are often infected, despite Maasai producers' conscious attempts to avoid pasture frequented by wildebeest during and following the calving season, which at present is the only practical way to prevent this disease [Mushi, Rurangirwa, and Karstad 1981].

Other diseases such as that caused by the beef tapeworm are also widespread among pastoral producers. The presence of Cysticercus bovis, the cystic stage of the beef tapeworm, Taenia saginata, is not dramatic and in fact cannot be verified until the time of the animal's slaughter by post mortem examination. Humans become carriers of this tapeworm by eating beef containing viable cysticerci. Cattle, in turn, are infected by eating grass which is contaminated by human feces containing the parasite's eggs. Consequently, restricting animal contact with human feces and killing the cysticerci by cooking or freezing meat are methods of breaking the cycle. The highest incidence of the disease in Kenya is found among lower-grade animals as shown by KMC's recorded cases during the mid-1970s (Table 3.10).

Heavily infested carcasses are condemned by KMC, while those lightly affected are simply frozen or boiled before sale. Economic losses arise from the outright condemnation of heavily infested carcasses and organs, the costs of treating lightly infested carcasses, and the loss in value of the latter carcasses following treatment [Grindle 1978]. The estimated losses to Kenya in the mid-1970s due to Cysticercus bovis, excluding human health costs, are shown in Table 3.11.

Even though the economic costs of diseases such as that caused by the beef tapeworm are great, the Kenyan Government has had to prioritize the use of the limited veterinary resources at its disposal by concentrating upon those health problems which present the greatest threat to livestock production and development. Among the major animal health concerns of the government are foot-and-mouth disease and tick-borne diseases, particularly East Coast fever. These diseases and measures taken to control them have significant direct impact upon Maasai producers, and are therefore discussed in greater detail.

Foot-and-mouth disease. Foot-and-mouth is a very contagious viral disease mainly affecting cloven-footed animals, both domestic and wild. It is characterized by fever and by vesicle and ulcer formation in the mouth, on the feet, and on the udder and teats of females. Of the seven types of virus recognized, all but two of them occur in Kenya. Mortality rates are high in young calves, and

Table 3.10. Incidence of _Cysticercus bovis_, by grade, recorded by the Kenya Meat Commission, 1974-76 average

Grade	Proportion of Total Head Slaughtered in Each Grade (1)	Incidence of Cysts by Grade (2)	Weighted Average of Incidence (1) x (2)
	-----------------	percent	-----------------
Prime	--	8.1 (Prime, Choice, FAQ)	1.6
Choice	7		
FAQ	13		
Standard	24	18.6	4.5
Commercial	52	25.2	13.1
Manufacturing	3	23.1	0.7
Condemned	1	--	--
Total			19.9

Source: Based on Grindle [1978], Table IV.

Table 3.11. Estimated cost of *Cysticercus bovis* to Kenyan abattoirs, mid-1970s

<table>
<tr><th>Source of Cost</th><th>Kenya Meat Commission</th><th>Local Abattoirs</th><th>Total</th></tr>
<tr><td></td><td colspan="3">----------------Ksh----------------</td></tr>
<tr><td>Condemnations</td><td>908,800</td><td>951,400</td><td>1,860,200</td></tr>
<tr><td>Extra costs, including treatment</td><td>951,400</td><td rowspan="2">10,834,600</td><td rowspan="2">11,984,800</td></tr>
<tr><td>Loss in value</td><td>198,800</td></tr>
<tr><td>Total</td><td>2,059,000</td><td>11,786,000</td><td>13,845,000</td></tr>
</table>

Source: Based on Grindle [1978], Table VI.

though low for adults, infertility among breeding animals commonly occurs. In a study which estimated the impact over three years of a foot-and-mouth disease outbreak on a herd of 2,000 head in Kenya, it was calculated that a minimum 10 percent reduction in meat production would result [Chemonics International Consulting Division 1977].

Mass, periodic vaccination has been the strategy pursued by Kenya's Department of Veterinary Services in its Foot-and-Mouth Disease Control Project. Compulsory foot-and-mouth vaccinations were initiated on an experimental basis in 1966, on a few ranches surrounding KMC's Athi River abattoir [Ngulo 1978]. Expansion of the program has proceeded, and today the zone of vaccination covers Maasailand as well as almost all higher-potential areas of the country. Nationwide coverage is the long-term goal [Kenya, Republic of 1980a]. If an outbreak occurs, the infected area is placed under quarantine, with movement of stock from or through the area prohibited. However, illegal movement of stock frequently takes place, frustrating the government's attempts to prevent transmission.

The number of cattle immunized under the compulsory vaccination program in Maasailand from 1975 to 1979 is shown in Table 3.12, along with the number of reported foot-and-mouth disease outbreaks. When treatment is postponed or delayed due to weather or shortage of veterinary staff, the disease can quickly resurface. For example, when the Maasailand campaign was disrupted in 1978 by heavy rains, there were subsequent outbreaks in the areas where vaccine had not been administered [Kenya, Department of Veterinary Services 1979]. Despite the government's considerable efforts, the disease seems at times intractable. Indicative of the insidious nature of the problem, an outbreak of foot-and-mouth disease in the vicinity of Nairobi prevented livestock exhibits at the 1982 International Nairobi Show, Kenya's largest agricultural fair.

East Coast fever. Tick-borne diseases such as redwater, heartwater and anaplasmosis occur in Kenya, but by far the most serious problem is East Coast fever (ECF). It is caused by protozoal *Theileria* parasites transmitted principally by the Brown Ear Tick, *Rhipicephalus appendiculatus*. Over 80 percent of Kenya's cattle share their environment with this tick, and over 50,000 head fall victim to ECF each year, making it the major cause of death among adult cattle in Kenya. Even so, Dolan and Young note: "It is possible that the economic losses due to chronic theileriosis are as great or greater than those caused by mortality" [1981, p. 413]. As with most livestock diseases, ECF becomes more acute when the animal is under stress. During the 1976 drought, when Kajiado District lost over half of its cattle population, ECF was reportedly responsible for a large share of the deaths [Kenya, Department of Veterinary Services 1977].

Table 3.12. Foot-and-mouth vaccinations administered and reported foot-and-mouth disease outbreaks, Maasailand, 1975-1979

	Kajiado District		Narok District	
Year	Vaccinations	Outbreaks	Vaccinations	Outbreaks
	---number---			
1975	344,000	53	789,924	23
1976	498,420	12	676,822	12
1977	451,126	6	495,846[a]	7
1978	349,517	14	1,015,977	--[b]
1979	798,819	10	819,163	7

Source: Kenya, Department of Veterinary Services [1976, 1977, 1978, 1979, 1980].

[a]First vaccination round only.

[b]Not reported.

No immunizing treatment for ECF exists, although the development of a vaccine using buffalo as a source of antigenic material is under current research [Grootenhuis and Young 1981]. Consequently, control of ECF depends on the control of ticks. To combat tick burden, livestock are dipped or sprayed using an appropriate acaricide, preferably at intervals of one week or less [Bram 1975; Drummond 1976]. The government has therefore concentrated on encouraging the construction, maintenance and use of spray races and especially dips.

The number of dips reported operating in Maasailand, 1975 to 1979, is shown in Table 3.13. The variability among the numbers for Narok District reflects the fact that the presence of a dip in an area and its operation are two very different conditions. In fact, dip maintenance and use have had an uneven history in Kenya. Livingstone [1975] found the condition for the country as a whole in the mid-1970s "atrocious" and, given the external diseconomies of nondipping, recommended that the government provide dipping services free of charge at least in regions strategic to livestock development. The government has not resorted to free dipping services, but is committed to assuming the management of dips if necessary [Kenya, Republic of 1980a].

Recently, there have been more favorable reports of producers' awareness of the importance of tick control, and an increasing willingness on their part to properly use and maintain dips. A project funded by the Danish aid organization, DANIDA, to establish a network of properly operated dips in Kericho District, illustrates the impact possible when a producer participatory approach is taken. Only one year after the project began, the number of dips operating in the district had risen from 68 to 290, and the average number of cattle dipped weekly increased from 79,246 to 416,250. Livestock producers contribute their own labor for dip construction and repair. A sense of responsibility instilled among users regarding dip maintenance has been identified as a principal reason for the project's record of success [Christensen 1980]. Recognition of the advantages of maintaining and using dips has also been reported for the Maasai. White and Meadows [1981] describe not only the regular dipping of livestock on Maasai ranches provided with dips, but also an appreciation for the benefits of doing so among producers on ranches without dipping facilities.

Control of disease is essential for industry stratification and the pastoral transition in Maasailand. This overview of disease control measures has highlighted both the importance and complexity of this area of governmental intervention. Ultimately, effective disease control requires conscientious management, and its benefits depend upon complementary activities in other areas of livestock development. The importance of such managerial interrelationships is exemplified in the

Table 3.13. Cattle dips in Maasailand in operation by yearend, 1975-1979

Year	Kajiado District	Narok District
	---number---	
1975	142	41
1976	149	41
1977	153	40
1978	157	21
1979	158	65

Source: Kenya, Department of Veterinary Services [1976, 1977, 1978, 1979, 1980].

following discussion on breeding interventions, which emphasizes the necessity of disease control and related practices for the successful introduction and maintenance of genetic improvements.

Upgrading Livestock

Crossbreeding. The majority of the cattle in Kenya belong to two basic strains of shorthorn zebu, the Large East African Zebu, of which the Boran breed is most prominent, and the Small East African Zebu, as typified by the cattle of the Maasai [Dahl and Hjort 1976]. The limited genetic potential of the indigenous zebu is often identified as a major constraint to beef and milk production, and thus breeding programs are a principal form of intervention [Trail 1981]. The greater advancements in the upgrading of herds have occurred in the higher-potential areas of Kenya, while ecological and disease factors have limited the extension of breeding programs to rangeland regions like Maasailand [Callow 1978; Griffen and Allonby 1979a].

Breeding policies which have been suggested for the various eco-climatic regions of Kenya are shown in Table 3.14. In the higher-potential Zones II and III, land is the factor of least elastic supply, and only very productive cattle such as European dairy breeds are economically competitive. In the drier Zones IV and V, cattle crossbred between European dual-purpose breeds (Friesian, Simmental, Brown Swiss, Red Poll) and the Sahiwal have been identified as likely to make best use of the environment [Meyn and Wilkins 1974]. In the very arid parts of Kenya (Zone VI), Sahiwal-zebu crosses are a possibility, but even then the productive potential is not great. Regardless of ecological conditions, management is the key element to successful breeding interventions, and this fact underlies the different upgrading policies for subsistence and commercial herds suggested by Meyn and Mbogo [1976] (Table 3.14). As reported by Semenye with regard to herds on a group ranch in Maasailand: "A comparison of measurements and weights between Zebu and cross-bred cattle at Elangata Wuas group ranch suggests that crossbreeding is not likely to increase productivity significantly unless it is accompanied by improved management" [1980, p. 27].

Where management practices have been suitable, the Sahiwal breed has had a significant impact on production [Trail and Gregory 1981a]. A dairy breed first introduced from India and Pakistan, the Sahiwal in Kenya usefully fulfills a dual-purpose role especially appropriate for pastoral producers in transition, such as the Maasai [Preston 1977]. Establishment of the National Sahiwal Stud in Kenya in 1962 marked the beginning of a program to produce this dual-purpose *Bos indicus* breed, intended for extensive pasture conditions with minimal supplemental

Table 3.14. Eco-climatic conditions and suggested cattle breeding policies for subsistence and commercial livestock producers in Kenya

	Breeding Policy	
Eco-Climatic Zone	Subsistence Producers	Commercial Producers
II, III	European dairy breeds	European dairy, dual purpose and beef breeds
IV	European x zebu dual purpose crosses	European x zebu beef crosses
V	Improved dual purpose zebu (Sahiwal)	Improved beef zebu (Santa Gertrudis)
VI	Indigenous	--

Source: Meyn and Mbogo [1976], Table 6.3.4.

feeding [Trail and Gregory 1981b]. In a study which compared the productivity of Sahiwal and other breeds under various management and ecological conditions, Trail and Gregory concluded that "the marked superiority of the Sahiwal to the Small East African Zebu, and its similar productive capacity to the Boran, show that it is well adapted for use to produce both beef and milk in [lower-potential] beef production environments" [1981b, p. 68]. The productivity values supporting this conclusion are shown in Table 3.15.

The Sahiwal may also possess a genetic tendency toward trypanotolerance. Data gathered over six years on a breeding herd at a dairy ranch near Kenya's coast indicated that in the 13 months following parturition the two-thirds Sahiwal, one-third Ayrshire crosses required less than half the number of treatments for trypanosomiasis as did the one-third Sahiwal, two-thirds Ayrshire crosses [Murray et al. 1981].

Artificial insemination. Provision of artificial insemination (AI) services is also an important area of governmental intervention for upgrading herds. As with crossbreeding, AI services have focused upon the areas where economic returns are greatest due to ecological and managerial conditions--the higher-potential areas of the country. An AI service to farmers began in Kenya in 1935, when the principal objective was the eradication of breeding diseases from settlers' herds. Today, AI services in these areas relieve smallholders of the necessity of keeping breeding bulls and give them access to genetically superior sires.

While the government's stated policy is to extend AI services to all parts of the country where there is potential for improved livestock production [Kenya, Republic of 1980a], concentration on *Bos taurus* breeds and cross breeds in the higher-potential areas remains considerable. The total population of female cattle in the country with more than 50 percent *Bos taurus* blood was approximately 500,000 in 1975, and of these an estimated 277,000 were covered by national AI services. In contrast, only about 45,000 zebu females were reported as normally presented for service annually [Duncanson 1975]. Infrastructural underdevelopment and unsuitable managerial practices prevent wider use of AI services in the rangelands [Meyn and Mbogo 1976].

Realization by Maasai producers of the economic benefits of herd upgrading, as with disease control measures, depends upon the management variable. An important influence upon management is exerted by the cohabitants and co-users of the range, the herds of wild ungulates and other game. Existing and potential wildlife management interventions are the following subjects addressed.

Table 3.15. Productivity values for the Small East African Zebu and Sahiwal breeds in a low rainfall, low and highly variable nutritive environment

Item	Unit	Small East African Zebu	Sahiwal
Cow viability	%	97	94
Calving percentage	%	74	80
Calf survival	%	95	96
Calf weight at one year	kg	128	152
Lactation yield	kg	34	595
Productivity index[a] per cow per year	kg	91	160
Cow weight	kg	280[b]	400[b]
Productivity index[a] per 100 kg of cow weight maintained per year	kg	32[b]	40[b]

Source: Trail and Gregory [1981b], Table 2.

[a]Total weight of one-year-old calf plus liveweight equivalent of milk produced.

[b]Estimated.

Managing Wildlife

Kenya's wildlife resources constitute an increasingly important sector of the nation's economy. Governmental wildlife interventions directly affect Maasai livestock producers and influence the transition process, since Maasailand is the home of world-renowned game parks. Consequently, planning the use of wildlife resources is a process impinging upon livestock development. As Heady observed shortly before Kenya's independence, "in no other aspect of managing the land resources in East Africa is so little unbiased biological information available and so badly needed" [1960, p. 113]. The knowledge base has expanded over the past 20 years, but the essence of this statement still holds true.

The government's commitment to wildlife development derives from a perspective which views these resources as yielding higher economic returns, if properly managed, than any feasible alternative use of the country's rangelands [Kenya, Ministry of Economic Planning and Community Affairs 1979a]. There are two aspects to this optimistic perspective: utilization of wildlife as food, and the economic gains obtainable through wildlife-based tourism.

Wildlife is an important source of protein for human consumption in much of Subsaharan Africa even though little effort has been made toward its management as a food source [de Vos 1977]. The greater efficiency in meat production of wild ungulates over cattle has been repeatedly noted in certain environments [Hopcraft 1975; Monod 1975; Reul 1979], and experiments in Kenya with game ranching are now well underway. But more immediate and dramatic in economic benefit have been the national parks and game reserves, which helped to attract over 300,000 vacationing foreign visitors to Kenya in 1979 [Kenya, Ministry of Economic Planning and Development 1980a]. At that time, tourism was second only to coffee as the country's leading foreign exchange earner [Ambrose 1980]. An estimated 25 percent of total foreign exchange earnings from tourism comes from wildlife-based tourism activities, and another 25 percent is due to visitors whose main activity is other than viewing, but who spend money on visits to game parks as well [Richards 1980].

Before the revenue from game viewing (and from hunting until it was banned in Kenya in 1977) and the potential for returns from game ranching made evident the benefits of wildlife to the economy, wild herds were often viewed by European ranchers as simply grazing competitors and transmitters of disease. General condemnation has given way to a more balanced assessment of the benefits and costs associated with wild herds, which recognizes that the degree of competition between wildlife and domestic stock is largely dependent on the species involved, the state of ranch development, and intensity of stocking [Casebeer and Denny 1967]. Game ranching capitalizes on the advantages of wild ungulates over cattle, such as their biological efficiency, greater disease tolerance, and more effective use of the plant and soil environments of semiarid and arid lands [King and Heath 1975; Surujbally 1977].

The major game ranching effort in Kenya is the Galana Game Ranch, established on land leased from the government between Tsavo National Park and the coastal belt, where buffalo, eland, and oryx are produced in addition to Boran cattle and small stock. Table 3.16 compares the returns to the Galana Game Ranch of cattle, sheep and oryx sold to a coastal tourist resort in 1975. Thresher [1980], employing a budgeting approach, assessed the relative financial merits of ranching oryx and Boran cattle at Galana using production coefficients as shown in Table 3.17. He computed the before-tax net profit, after assessing a 10 percent return on stock and land market values, as Ksh 140,000 for oryx in contrast to a negative Ksh 450,000 for a breeding herd of Boran cattle.

Game ranching ideally aims at a mix of species rather than complete replacement of domestic animals by wild herds. Eland, for example, can be effectively herded together with domestic herds, since the eland's principal browse diet complements cattle grazing. However, the successful introduction of commercial game ranching in Maasailand, while a productively attractive possibility, would depend upon producers' acceptance and a sufficiently high level of demand for the game meat once it is produced. To convince Maasai to include wildlife in their herds and develop proper management programs would be extremely difficult given social and cultural orientations [Jaffe 1975; Maloiy and Heady 1965]. Moreover, the meat of game has tended to remain a specialty dish mainly marketed to restaurants and tourist lodges. Higher processing costs per kg for game meat as compared with cattle also limits production possibilities, since cost per head slaughtered is essentially the same for all species [McDowell 1984]. In some Subsaharan African countries, such as Botswana and Zaire, most of the meat consumed is game [Krostitz 1979]. But the consumption level for Kenyans, estimated at 0.5 kg per capita per year, is negligible, even though "popularization of its consumption" is a stated government objective [Kenya, Republic of 1980a, p. 24].

As mentioned, the revenue earning capacity of game is much more evident in the growth of Kenya's tourist industry, and it is this aspect of wildlife resource use which more immediately affects Maasai livestock producers. In Chapter II, reference was made to the encapsulation of Maasailand which has proceeded through the formation and expansion of game parks. In the past, benefits from the wildlife-based tourist industry were split between private entrepreneurs and the government, while costs in terms of

Table 3.16. Value to the Galana Game Ranch of cattle, oryx, and sheep sold to a coastal tourist resort, 1975

	Cattle	Oryx	Sheep
	----------------	Ksh	----------------
Price per kg	6	8	6
Value of			
carcass[a]	1,100	656	108
hide	--	100	--
head	--	60	--
Total	1,100	816	108
Value per ranch unit[b]	1,100	1,632	540

Source: King and Heath [1975], Table 4.

[a]Dressed carcass weights: cattle, 183 kg; oryx, 82 kg; sheep, 18 kg.

[b]Number of head per ranch unit: cattle, one; oryx, two; sheep, five.

Table 3.17. Production coefficients and potential offtake levels for oryx and Boran cattle, Galana Game Ranch

Item	Unit	Oryx	Boran Cattle
Useful life	year	10.5	12
First calving	year	2.5	3.5
Breeding	% per year	112	75
Weaning	% per year	100	70
Slaughter age	year	2.0	3.75
Liveweight	kg	135	350
Annual offtake	%	30	20
	head	3,300[a]	1,000[a]
Mortality loss[b]	head	80	25
Saleable animals	head	3,220	975
Saleable liveweight	kg	434,700	341,250

Source: Based on Thresher [1980], Table 1.

[a]Assuming 11,000 oryx or 5,000 Boran cattle, for an animal biomass of 22 kg per ha, a reasonable grazing pressure in Galana-type country.

[b]Assuming 2.5 percent mortality rate.

reduced grazing and watering resources were borne by the Maasai. Presently, in accordance with the alleviation-of-poverty theme of Kenya's Development Plan, 1979-83, increased attention is being paid to the equitable distribution of wildlife-derived benefits.

> The main objective of tourism and wildlife development is to maximize net returns to investments, subject to important social and environmental constraints. In the context of the Development Plan 1979-83 "net returns" is to be interpreted to include the sector's contribution to rural development and prosperity, especially in terms of income earning opportunities and provision of basic needs. [Kenya, Ministry of Economic Planning and Community Affairs 1979a, p. 95]

Pilot studies have been carried out as a part of the government's Wildlife and Tourism Project, with the objective of reconciling conservation and development aims, local and national needs, and wildlife-based and alternative uses of the land. Two of the three study areas, Masai Mara and Amboseli, lie within Maasailand. Though not specifically targeting livestock production like the other areas of intervention which have been discussed, government wildlife policies directly affect Maasai livestock producers. In the short run, issues related to benefits and costs associated with the operation of game parks are the major area of impact. For example, visits to the Amboseli Game Reserve in Kajiado District have increased from 10 to 25 percent per year since 1967, and Thresher [1981] estimates that one Amboseli maned lion has a present value to the national economy equivalent to a base herd of 30,000 zebu cattle or an annual offtake of 6,400 steers. A growing source of foreign exchange of this magnitude will necessarily have a major influence on regional development planning. In the longer run, the management of game in association with cattle and small stock could well become a source of increased production and therefore figure significantly in the transition process.

Summary

Several interacting facets of governmental activity affecting the pastoral transition have been examined in this section. The discussion of livestock diseases centered mainly on foot-and-mouth disease, against which the government has mounted a long-term campaign involving Maasailand, and East Coast fever, the major killer of

Maasai stock. Livestock breeding interventions were considered with respect to the upgrading of indigenous herds, for which the Sahiwal breed has proven especially appropriate for pastoral areas, and constraints to the expanded use of artificial insemination. Finally, the discussion of wildlife interventions has shown the government's awareness of the need to include the wildlife variable in planning for the optimal use of Maasailand's resources. Measures taken to control disease, upgrade stock, and manage wildlife resources are necessarily interrelated, an essential fact in the planning of successful livestock interventions in Maasailand.

CONCLUSIONS

Firey [1960] notes that if resource plans and policies are to be more than simply opportunistic ventures, they must be part of a logically closed system of thought that embraces the whole range of governmental activity. The truth contained in Firey's observation has become apparent in this discussion of governmental activities which influence both the transition process in Maasailand and, more generally, the stratification of Kenya's livestock industry. Policies and programs which contradict these concomitant regional and national goals have been recounted, as have fields of intervention which contribute positively to expanded livestock production.

Pastoralists' preoccupation with breeding stock is likely to facilitate the transition process, since the relatively intensive level of supervision of animals demanded in cow-calf operations is already present [Hjort 1981; Pratt and Gwynne 1977; von Kaufmann 1976]. But the process is hindered by institutional obstacles. Administered meat prices, perhaps the most pervasive form of intervention, exemplify the negative impact on production of short-sighted policies motivated by near-term political advantage. Meat price controls, justified as contributing to the poverty alleviation theme of the Development Plan, 1979-83, are officially viewed as an attempt to mediate fairly between the interests of producers and consumers. But, as shown, meat price controls have severely retarded the stratification and transition processes. An unrealistic, unresponsive price system has meant little mobilization of resources to increase output, an especially serious situation since the most important long-run effects of price incentives on production are through price-induced shifts in the production functions [Brown 1978]. As it is, price incentives received by Kenya's Maasai and other producers have hardly been sufficient for maintaining existing production levels, let alone prompt the adoption of new methods and technologies. Yet it has long been recognized that a policy promoting higher producer prices would be a powerful developmental tool for the livestock industry [Schaefer-Kehnert 1968]. A price system based upon market supply and demand is not perfect but, as noted by Samuelson [1973] and exemplified in Kenya, neither is one of administered control.

Descriptions of the eclipse of KMC's operations by private abattoirs and the demise of feedlotting enterprises during the 1970s illustrate the essential role played by prices and price ratios in generating and channeling production. Relative increases in meat prices in the early 1980s suggest that the government is becoming aware of the crucial importance of favorable prices to industry stratification and the pastoral transition.

In spite of biased meat pricing policies, the internal stratification of rangeland production, with immatures from the northern pastoral areas grazed on ranches in Maasailand, is expanding, encouraged by the government's steer fattening loan program. The expansion of steer feeding, despite the low rates of gain possible on unimproved range, is indicative of the responsiveness of Maasai producers to profit-making opportunities. In this case, the government's interventions are affecting positively the transition process.

Regarding veterinary interventions, the emphasis given preventive treatment by the government, as in compulsory foot-and-mouth vaccination campaigns and the promotion of tick control programs, has improved the health of Maasai herds and flocks. With increased research into disease-environment relationships, and especially disease problems prevalent in pastoral livestock systems, the effectiveness of disease control measures is likely to increase accordingly [Awogbade 1979; Griffen and Allonby 1979b]. Less obvious than for marketing and disease control interventions is the impact of breeding programs on stratification and the transition process. Years are required before results are realized, and even then depend critically upon livestock management levels.

Finally, the immense value of the game parks to the national economy and the potential gains to be derived from game ranching, as an alternative use of pastoral lands, add another dimension to the transition process. As the pastoral transition in Maasailand proceeds, and resource use is more carefully monitored and allocated, the roles of wildlife as competing and complementing resource converters will require increasingly precise identification. The economic importance of game parks will continue to grow over time, but whether game ranching is one day incorporated into livestock operations will ultimately depend upon production costs and demand preferences.

The government is pursuing a rational course, developing its tourist industry while promoting experimentation with game ranching as an alternative use of range resources. Still, the issues of equitable distribution of the benefits and costs regarding wildlife resources and the physical impact on Maasai producers become evermore complex as land use pressures intensify. In the long run, the production of game meat may well serve both economic and ecological interests for, as Howard notes, "when the fauna of this country is offered to private enterprise for commercial exploitation the conservation of wildlife is ensured" [1981, p. 67].

No concluding statement wholly laudatory or entirely unfavorable can be made regarding the impact of existing governmental interventions upon pastoral production. Other than for meat pricing policies, the various interventions discussed are directed toward objectives in accordance with the stratification and transition goals. But for the interventions to effectively facilitate the attainment of these goals, a receptive and progressive managerial base is necessary. Buck et al. [1982] have shown in Botswana that improved performance due to breeding interventions is dependent upon the management factor. Breeding activities and managerial practices in general, especially disease control, collectively build one upon the other. For example, strict disease control and a high standard of animal husbandry are necessary for AI services to be successful in the upgrading of herds [Duncanson 1975]. Likewise, in the administering of disease control measures, Thompson et al. caution that "all acaricide and anthelmintic treatment should be used only as an aid to good managerial practices" [1978, p. 144].

If, as noted by Wyeth [1981], pastoral producers such as the Maasai, especially those who live in areas unsuitable for cultivation, want to raise the productivity of their animals before all else, then the end goals of the government (the nation) and of producers are the same [Simpson 1983]. However, political and institutional constraints such as meat price controls deflect both parties from a concerted effort to attain this goal. Ultimately, producers' practices and objectives determine the rate at which the transition process will occur. In order to understand the impact of price controls and other constraints at the producer level, knowledge of Maasai livestock management activities and changing tenurial circumstances is necessary. They are the subjects of the following chapter.

4 The Maasai Pastoral Economy and Tenurial Change

This chapter provides a general statement on the livestock economy of the Maasai and the forms of tenurial intervention in the region. General characteristics of the society and principal livestock management practices, including marketing activities, are described. This overview serves as a basis for discussing the impact of tenurial change upon the traditional economy, as well as a point of departure for the descriptions of producer samples in Chapter V. The examination of tenurial interventions centers primarily on the group ranch, with an assessment of its achievements and shortcomings in furthering the development of the Maasai economy.

MAASAI PASTORAL ECONOMY

Reference is made repeatedly in the following discussion of the Maasai pastoral economy to one group ranch, Elangata Wuas, studied by the staff of ILCA's Kenya Country Programme [Bille and Anderson 1980; Eidheim and Wilson 1979; Semenye 1980; and Wilson, Peacock, and Sayers 1981]. These and other cited studies are illustrative, intended to provide an understanding of the magnitudes of pastoral parameters found in Maasailand.[1] No regional statistical representativeness should be assumed. Also, it should be noted that the discussion focuses on Kajiado District, with Narok District receiving relatively slight attention. This imbalance reflects the comparatively little research that has been undertaken in Narok and the paucity of information on the district. In areas where there are significant differences between the districts, the distinctions are duly noted.

Polity and Society

Political and territorial divisions. The Maasai were never organized as a single tribe under a unified political system. Rather, they belong to about a dozen "sections," or iloshon (sing. olosho), a division distinct from the Maasai lineage system of clans. Each olosho has its own territory, variation of Maasai customs and, traditionally, decision-making autonomy [Allan 1965]. The Maasai maintain strong social and political ties with their particular section, and the heads of households are entitled to grazing and watering rights within its boundaries. Historically, the institutionalized sharing of resources among iloshon such as during times of drought was common, but households were prepared to defend their section's boundaries, by force if necessary, against unauthorized intrusions by other Maasai [Jacobs 1975].

An olosho is further subdivided into inkutot (sing. enkutoto). "This term embraces a variety of local organizational arrangements in Maasai society" [Eidheim and Wilson 1979, Annex II, p. 5], but the defining trait apparently is the productive unity of a locality. The households of an enkutoto have access to a permanent water source and dry and wet season grazing lands [Jacobs 1963]. As self-contained ecological units, inkutot have also traditionally been the loci of local-level political activity, with each one having its own council of elders to manage affairs in the area. As the pastoral economy of Maasailand has come under increased population pressure, significance of the enkutoto has not diminished. The social and ecological appropriateness of currently adjudicated tenurial divisions is invariably measured against that of the traditional territorial units.

The age-set system. Underlying the territorial divisions, the Maasai share a common social structure based upon an age-set system.

> The life of a Maasai male is a well-ordered progression through a series of life-stages, which are determined by age, initiated through ceremonies, and marked by specific duties and privileges. The males of every Maasai section pass through three main stages: boyhood, warriorhood, and elderhood. Warriors are subdivided into junior and senior warriors and together form one generation or age-set. Approximately every fifteen years, a new generation of warriors comes of age. . . . When warriors graduate into elderhood, they are replaced by another generation of warriors. Elders progress through junior and then senior elderhood, and eventually become ancient elders, who, because of their old age, retire from the active direction of Maasai affairs. [Saitoti 1980, p. 30]

The age-set system of the Maasai underpins and is reinforced by their livestock-centric economy. Livestock management roles and responsibilities, and conditions by which livestock are accumulated, are broadly delimited by one's age set. The initiation into a new age set and other, less major ceremonies of social passage are celebrated by ritual slaughter and often entail dietary obligations. Hence, the social and economic lives of a male are woven together as he passes from boyhood (I'laiyok), to warriorhood (Ilmurran), to elderhood (Ilmoruak), and finally to ancient elderhood (Ildasati). Although women are not grouped into corporate age sets like men, they are identified with the male age set for whom they sang when young and unmarried.

Settlement and family organization. Most pastoralists live in relatively small, semi-permanent settlements called inkan'gitie (sing. enkang), generally thorn bush fenced enclosures commonly referred to as bomas, with individual homes located inside on the periphery. The center area is used to corral livestock at night. Most inkan'gitie are occupied by more than one household. A household, or olmarei (pl. ilmarei), is generally composed of a man, his wife or wives (the Maasai are polygynous), other adult dependents, and his unmarried children, although there are many variations to this pattern. The average household composition found on Elangata Wuas group ranch shown in Table 4.1, indicates that the 236 households of this group ranch were distributed among 64 inkan'gitie, with households per enkang ranging from 1 to 13 [Bille and Anderson 1980]. The households of an enkang may be related through extended family ties, but often are not.

Households are divided into subhouseholds, or nkajijik (sing. enkaji). Each subhousehold is usually under the control of a wife of the head of the household, and harbors a degree of economic independence. While a household's livestock do not graze as subhousehold units and are not corraled separately within the enkang at night, each wife is responsible for the milking and care of the particular animals belonging to her subhousehold. Decisions regarding the disposition of livestock ultimately belong to the head of the household, but ownership rights to most of the household's animals are allocated among wives and other dependents. Hence, the household and subhousehold are the social units within which most decisions regarding livestock management are made and carried out. Some of the management practices commonly found in Maasailand are summarized next.

Livestock Management Practices

Herding and related activities. The herding of cattle and small stock is generally assigned to children, although other members (and nonmembers) of the household may be involved, depending largely upon the availability of labor within the enkang. When animals are watered and

Table 4.1. Average household composition, Elangata Wuas group ranch, Kajiado District, 1980

Household Member	Mean Number	Proportion of Total
		--percent--
Head of household	1.00	10.6
Wives	1.42	15.1
Children under 15 years	4.24	45.1
Children over 15 years	0.38	4.0
Other dependents	2.34	24.0
Total	9.38	100.0

Source: Bille and Anderson [1980], Table 12.

dipped or sprayed to control ticks, older family members are likely to be in charge. Table 4.2 indicates these labor conditions for Elangata Wuas group ranch. Cattle are usually watered daily or every second day, a regime that varies seasonally. Small stock are often not watered as frequently as cattle. Watering patterns reported for Elangata Wuas group ranch are shown in Table 4.3.

A household's herds and flocks are often grazed together with those of other households, usually of the same enkang. Cows are milked by the women in the morning, before they are taken to graze for the day, and/or in the evening, after returning for the night. The threat of wild predators and thieves, not to mention the greater likelihood of stock simply becoming lost, make night grazing infeasible. Livestock theft occurs, though not as frequently as it once did. Losses to predators, on the other hand, are often substantial. As Wilson, Peacock, and Sayers report regarding the danger for small stock posed by jackels: "They may take stragglers that are separated from the flock, they also steal from the enclosure at night and may even steal lambs and kids close to the enclosure during the day" [1981, p. 13].

Disease control and breeding practices. Livestock disease control measures common in Maasailand include dipping or spraying cattle and sometimes small stock for ticks, usually on a weekly basis, drenching calves and small stock once or twice a year for internal parasites, and administering antibiotics, such as terramycin, to animals requiring treatment. Actual practices vary widely, with some households lax in their disease control practices and others more conscientious. Breeding controls entail the castration of most males, for breeding selection and to add to the value of animals destined for sale or home slaughter. Seasonally controlled breeding of small stock is attempted by some Maasai by attaching a plastic apron to the male during a rainy season. Wilson, Peacock, and Sayers [1981] found these attempts at control only partially successful and continuous breeding the norm, but Peacock [1982] reports effective control for particular localities, as described in Chapter V.

Livestock dispersion and livestock-based relationships. Animals belonging to a household are commonly distributed among relatives and friends who live at different inkan'gitie, a practice, as described in Chapter II, intended to reduce disease and climatic risks. Hedlund [1971] estimates that 30 to 40 percent of a Maasai pastoralist's cattle are dispersed in this manner. On Elangata Wuas group ranch, 31 percent of the cattle, 12 percent of the sheep, and 14 percent of the goats, or 30 percent of all livestock units in the survey, were reported in an enkang different from that of the household that owned them [Bille and Anderson 1980].

In addition to the spatial dispersion of owned stock, noncommercial livestock transfers and transactions are

Table 4.2. Persons principally responsible for herding and watering livestock, Elangata Wuas group ranch, Kajiado District

	Proportion of Total Respondents			
	Herding		Watering	
Person Responsible	Cattle	Small Stock	Cattle	Small Stock
	--------	--------	percent	--------
Head of household	1.3	0.9	9.7	10.5
Sons over 15 years	1.3	0.9	64.3	65.6
Children	78.5	77.9	8.8	8.0
Hired herdsmen	11.9	13.1	5.1	4.5
Combinations of above	7.0	7.2	12.1	11.5

Source: Bille and Anderson [1980], Tables 16 and 17.

Table 4.3. Watering frequency in wet and dry season, Elangata Wuas group ranch, Kajiado District

	Proportion of Total Respondents			
	Cattle		Small Stock	
Watering Frequency	Wet Season	Dry Season	Wet Season	Dry Season
	---------------percent--------------			
Without restriction	98.7	3.9	29.3	1.8
Every two days	1.3	93.0	19.5	63.7
Every three days	--	2.2	7.9	22.1
Less than every three days	--	0.9	43.3	12.1

Source: Bille and Anderson [1980], Table 18.

frequent. Their importance to the maintenance of the pastoral economy becomes apparent when disparities in livestock holdings are considered, as illustrated by the considerable range in size of herds and flocks reported for Elangata Wuas group ranch households (Table 4.4). Of the 236 households surveyed, 12 owned no livestock, and mean numbers of cattle, sheep, and goats were about twice the median holdings for each of the species [Bille and Anderson 1980]. The skewed distribution indicated by these data is supported by Peberdy:

> A recent survey in a particular section of Kenya Masai has shown that although the total cattle population is sufficient to provide subsistence, over 30 percent of the families have insufficient stock of their own to provide subsistence, and this is increased to 50 percent if their needs for cash to pay for taxes, school fees etc, are included. [1969, p. 166]

Dispersion of owned stock enhances a household's viability, and at the interhousehold level, livestock tranfers and transactions temper inequalities while giving emotional content and legal validity to relations between individuals and between groups [Eidheim and Wilson 1979]. As stated by Galaty, "rights in productive resources are social and are invariably constituted through exchange" [1981d, p. 71]. The pastoral economy is also reflected in dietary patterns. In the following sub-section, the predominantly livestock-based diet of the Maasai, the herd and flock structures implicit to this diet, and changing consumption patterns are addressed.

Food Production and Consumption

Diet and herd and flock structures. Historically, the dietary dependence of the Maasai upon their livestock was virtually complete, other than during periods of severe drought or other catastrophe. Even in recent times, it has been estimated that cow's milk constitutes as much as 80 percent of the Maasai pastoralist's diet, with meat, especially that of small stock, comprising a major portion of the remaining share [Unesco/UNEP/FAO 1979]. This extreme commitment to an animal-based diet, characterized even for a pastoral people as atypical [Jacobs 1975], is undergoing rapid change. While the Maasai of some sections and localities remain more strictly bound to a milk-and-meat diet than others, consumption of increasing levels of nonpastoral foods is the trend. This trend is due to a general decline in the size of livestock holdings and more frequent and regular contact with non-Maasai

Table 4.4. Distribution of livestock holdings belonging to member households of Elangata Wuas group ranch, Kajiado District

Quartile of Households[a]	Livestock Units[b]	Average Number of Livestock per Household: Cattle	Average Number of Livestock per Household: Small Stock
		----------head----------	
Poorest	<15	6	9
Second	15-35	20	20
Third	35-90	51	50
Wealthiest	>90	203	123

Source: Bille and Anderson [1980].

[a]Households numbered 236, with a total population of 2,214.

[b]One livestock unit = 1 bovine = 10 small stock.

societies [de Souza 1980]. However, despite altered dietary patterns, the Maasai above all still value their cattle for producing milk, and their small stock for supplying meat and fat [Bartenge 1980].

This valuation is reflected in the predominance of females in Maasai herd and flock structures, as exemplified in Table 4.5. Ideally, a subhousehold has cows lactating, and thus a supply of milk, at all times of the year. Eidheim and Wilson [1979] report normal lactation periods at times of very good range conditions of from four to seven months, with milk yields of 2 to 3 liters a day one to two months after calving. However, milk yields per cow are likely to drop to but a fraction of a liter per day when forage is scarce. The preponderance of cows in a typical herd makes a sufficient supply of milk throughout the year more likely, but the attainment of this ideal state remains subject to the high seasonal variability of yields.

Cow's milk is used to prepare sour milk and butter when forage is abundant and milk plentiful. When milk yields are low, consumption may be supplemented with goat's milk [Jacobs 1963]. White and Meadows [1981], in their 12-month study of group ranch households in Kajiado District, recorded only two occasions of goats being milked, in one instance when rains failed and another time when maize was in short supply due to a national shortage.

The slaughter and consumption of cattle, usually steers, is invariably for ceremonies, unless the animal is injured or accidentally dies. Small stock, on the other hand, are consumed more casually and frequently, though they may be slaughtered for special occasions as well. It is not uncommon for a sheep or goat to be slaughtered to make soup for someone ill or feeling weak, and the fat of sheep is considered beneficial for a woman following childbirth. Maasai tend to prefer goat meat to mutton, but at the same time they greatly enjoy the fat of their fat-tailed and fat-rumped sheep. It has been estimated that for every sheep and goat "officially" slaughtered in the rangelands of Subsaharan Africa, there are ten unrecorded slaughters [Wilson 1982]. The number of small stock slaughtered for home consumption in Maasailand, if not of this magnitude, is at least several times greater than the number sold to butchers.

Changing dietary patterns. Sugar and tea have long found a place in the Maasai diet. Milky, heavily sugared tea is the most common of prepared beverages. Maizemeal, which is eaten as a hard or soft porridge, and cooking fat are also common food purchases. The demand for these purchased substitutes for milk and milk products fluctuates seasonally. For example, Meadows and White [1979] found that the purchase of fats in Kajiado District declined by as much as 40 percent when milk was plentiful. However, maizemeal and cooking fats are increasingly being

Table 4.5. Cattle, sheep, and goat herd structures typical of Maasailand, 1980

Sex	Species		
	Cattle	Sheep	Goats
	------------percent-----------		
Female, breeding	41	44	48
nonbreeding	33	25	18
Male	10	16	24
Male castrate	16	15	10

Sources: Cattle--Based on Meadows and White [1979] and White and Meadows [1980b].
Small stock--Based on Elangata Wuas group ranch surveys, Wilson, Peacock, and Sayers [1981].

purchased and consumed on a regular basis by Maasai households.

> Traditionally the Maasai ate maizemeal under stress and not by choice. However, there has been some change over the last ten years and we are told that women and children now eat maizemeal, at least once weekly from choice. The men, however, still prefer their traditional milk diet, and only supplement it when milk supplies are inadequate. [Meadows and White 1979, p. 14]

Less frequently purchased foods include beans, rice, and potatoes. These food items are served to guests at celebrations. Two items showing very rapid increases in demand among "food" purchases are soft drinks and beer. Meadows and White [1979] report that the sale of Coca-Cola products in Kajiado District rose by 205 percent during the period 1972 to 1978, and beer sales by more than 250 percent between 1971 and 1977. No other purchased food items exhibited such significant upward trends during this time. Thus, dietary patterns are widening, with the increased demand for purchased food items drawing forth livestock sales.

This discussion has centered upon the diet of Maasai living in semiarid regions. In areas of higher agricultural potential, where Maasai cultivate maize, beans, and other food crops, there is a greater acceptance of and reliance upon nonpastoral foods. Milk is still the preferred food in these areas, but its consumption may be of secondary value nutritionally.

Livestock Marketing

The extent to which local-level marketing constraints in Maasailand hinder expanded livestock production is addressed in Chapter VII. Here, as part of the overview of the pastoral economy, producers' involvement in markets is examined in terms of offtake rates and principal determinants of the sale decision.

Over the 1960s and 1970s the Maasai have become increasingly involved in livestock sales, although records are grossly incomplete. Dispite the dearth of official information, it is widely recognized that the Maasai are cash conscious, and sell animals when it is to their perceived advantage. However, it would be as incorrect now to make a blank statement regarding willingness to market livestock as were colonial pronouncements to the opposite.

Pastoral offtake rates fluctuate widely over time, depending upon whether the period is one of relative growth, decline, or stability in herd and flock numbers. This long-term variation is apparent in estimations of the rates of offtake for cattle herds of Kajiado District from the 1960-61 drought to the late 1970s by Meadows and White [1979], as shown in Table 4.6. Following the drought, the offtake rate was estimated to have fallen to about 10 percent. As the recovery progressed, offtake rates increased to 13 and then to 16 percent, over a seven-year period. The next few years found herd growth rates slowing from an estimated 8 percent to 5 percent a year, with the rate of offtake about 17 percent. In 1973/74, as drought conditions developed, rates of offtake increased to 22 percent, and in the following two years, jumped to 35 and then to 38 percent, as herd numbers fell from an estimated 800,000 in 1973/74, to 500,000 in 1976/77. Following the drought, herd numbers began to recover, increasing to 540,000 in 1977/78, and the rate of offtake returned to a 13 percent level. This fluctuating record suggests that offtake rates in Maasailand are influenced primarily by range conditions.

Meadows and White [1979], in an effort to more fully explore this and other determinants of cattle sales in Kajiado District, examined relationships between the rate of offtake and (a) rainfall, (b) availability of grown steers, (c) incidence of foot-and-mouth disease, and (d) cash needs. They found a negative correlation of 0.36 between annual rainfall and the level of cattle sales in the district over the period 1956 to 1977, indicating that with improved range conditions, the supply of livestock to markets declines. The unavailability of grown steers apparently limited sales in the 1960s more than in the 1970s since, during the latter decade, Maasai were increasingly willing to sell not only younger steers, but also females, mainly culled cows and heifers not calving. Foot-and-mouth disease, specifically, the local quarantines imposed during outbreaks of the disease, was thought likely to restrict sales. But Meadows and White [1979] concluded that outbreaks were not a major constraint to cattle marketing in Kajiado, since sales over the period 1960-77 were at their highest when the incidence of foot-and-mouth disease was also high, from 1973 to 1975. Finally, the increase in cash demands principally resulting from the changing diet of the Maasai was found to be a major determinant of cattle sales, despite the fact that the effect of seasonal fluctuations in milk supplies on offtake could not be detected. This latter unexpected result was attributed to imprecise measurement of domestic milk supplies, and has been contradicted by more rigorous examinations of livestock sales and consumption expenditures [Grandin and Bekure 1982; White and Meadows 1981].

Table 4.6. Estimated cattle population, offtake, and rate of offtake, Kajiado District, 1962/63-1977/78

Year	Cattle Population	Offtake	Rate of Offtake
	---head---	--head--	--percent--
1962/63	206,300	20,955	10.0
1963/64	298,800	40,974	13.7
1964/65	373,100	49,293	13.2
1965/66	428,700	50,805	11.8
1966/67	493,500	74,026	15.0
1967/68	533,300	83,343	15.6
1968/69	572,800	91,396	16.0
1969/70	621,400	103,197	16.6
1970/71	669,500	111,830	16.7
1971/72	719,600	119,887	16.7
1972/73	757,200	127,130	16.8
1973/74	798,700	175,731	22.0
1974/75	793,500	275,120	34.7
1975/76	651,100	249,866	38.4
1976/77	503,400	104,469	20.8
1977/78	547,700	73,761	13.5

Source: Meadows and White [1979].

Patterns of offtake vary among households. Konczacki among others has hypothesized that marketable surpluses are a function of the distributional patterns of livestock ownership: "Given the intensity of the desire for security provided by livestock, families with larger herds should be more willing to sell some of their animals than those which own herds of smaller size" [1978, p. 49]. Wealthier households do sell a larger absolute number of animals, but rate of offtake and herd size have been found to be inversely related [Bekure and Grandin 1982; Peberdy 1969]. Variations in offtake by species have also been discerned. White and Meadows, in their study of Kajiado group ranch households, found "an increasing dependence on income from cattle sales, the wealthier the household, and an increasing dependence on income from small stock sales the poorer the household" [1981, p. vi].

The relationship between social position and livestock sales is more complex than general correlations between stock wealth and level or type of commercial offtake might suggest. For example, on Elangata Wuas group ranch one-third of the poorest households sold relatively large numbers of animals while another third of the same wealth group did not sell any livestock [Bille and Anderson 1980]. Similarly, the offtake patterns of sampled producers discussed in Chapters V and VI vary widely. A society like that of the Maasai cannot be described in terms of a single mode of exchange [Kapferer 1976; Keesing 1976], and the prevalence of nonmarket transactions such as livestock exchanges, gifts, and loans also influences the frequency of sales. Sales can be expected to increase in comparison to these other forms of transaction as the Maasai become more dependent upon purchased goods. In sum, certain factors have a widespread influence on rates of commercial offtake, such as the weather (expected range conditions), while others, particularly immediate cash demands, result in short-term fluctuations in sale levels. Range conditions and herd size set general levels of market involvement, but the particular production circumstances which each pastoralist faces--and his perception of opportunities and costs--ultimately determine the sale decision.

Summary

Maasai social structures and practices have been described, and traditional norms have been shown to be one set of constraints among many that channel choice [Heath 1976]. Local political and social institutions, settlement organization, and livestock management practices characterize a mode of production well-suited for meeting subsistence objectives. But, customs and institutions are undergoing modification in the present environment of rapid demographic change. Increasing levels of market involvement as illustrated by changing dietary patterns permeate the economy. Such changes signaling the transition process, however, are often not readily accommodated by existing pastoral institutions.

In neighboring Tanzania'a Maasailand, Jacobs [1978] observes that besides a decline in externally generated improvements such as bush control and veterinary services over the past 20 years, there have been many subtle, internal socioeconomic changes: Young men of the warrior age set are routinely herding animals (usually childrens' responsibility), livestock management duties are being passed on at an earlier age, and there is a radical decrease in the average number of households per enkang. Moreover, customary management practices have fallen into disuse. "Local elders no longer seem to meet as they did 20 years ago as a `council of elders' (enkiguana) to decide on matters relating to good pasture management, or to censure and take action against those who were practicing bad management" [Jacobs 1978, p. 26].

An increasing laxness in livestock management practices has been observed in Kenya's Maasailand as well. Men are reportedly away from their households and livestock more frequently than in the past now that they have bicycles and increased access to public transport. Also, smaller herds have led to careless management since livestock are more likely to be left in the care of women and children with limited supervision than are larger herds [de Souza 1980]. But the growing casualness in livestock management practices is not recognized by the Maasai as having adversely affected the productiveness of their herds, due to the compensating impact of recently acquired watering and dipping facilities [de Souza 1980].

Demographic and economic trends are apparently inducing Maasai pastoralists to relax their hold on livestock management practices which they once grasped firmly, rather than prompting the adoption of increasingly productive activities. Changing consumer demands are drawing forth higher levels of offtake, but as long as strategies are derived within the pastoral framework and are governed by subsistence oriented objectives, the system's capacity for accommodating change is extremely limited [Jahnke 1982]. As was succinctly noted by the World Bank's KLDP II Review Mission: "Pressure to keep a high and even increased number of females in order to provide milk for subsistence will probably be the overriding concern from the Masai's point of view; under these circumstances the management alternatives are quite limited" [IBRD 1977, Annex 6, p. 37]. A key element in expanding the producer's set of management alternatives will be resource control; the short history of current tenurial changes taking place in Maasailand provides insight into the difficulties of establishing this control.

FORMS OF TENURIAL CHANGE

Group and Individual Ranches

Group ranches: conception and reception. Practically all of the communally grazed lands of Kajiado and Narok Districts were classified as trust land at the time of Kenya's independence. The group ranch, initiated as part of KLDP, was to be the principal entity by which these lands would be transformed into deeded holdings, with rights and responsibilities of land ownership devolving to specific household heads, namely, the group ranch members. Because this proposal for titled communal ownership was so unlike existing forms of land tenure, legislation had to be specially drafted and enacted in the form of the Land (Group Representatives) Act, 1968, and the Land Adjudication Act, 1968. These laws provided for the adjudication and registration of group rights to land, and the means whereby, through the election and incorporation of group representatives, ranch management and development activities could be undertaken.

The World Bank envisaged the group ranch as the structural form through which interrelated range management and economic development goals could be attained. Ideally, the group ranch would facilitate the shift from subsistence milk production to meat production for the market, by establishing the collateral conditions necessary for major loan investments. Land use would improve, since it would be in the interest of group members to limit stocking rates to a ranch's carrying capacity. Social objectives would be achieved, from a more equitable distribution of stock wealth and movement away from deep social interdependencies, to improvements in education and other social services which would afford the surplus population greater mobility and employment opportunities [IBRD 1977].

Unofficaliy, the group ranch concept was more honestly recognized as simply the least objectionable means by which to implement tenurial change in Maasailand [McCauley 1976]. As Helland noted, "the problems of human and livestock population growth, combined with finite (or even shrinking) range resources and leading to over-grazing and environmental deterioration, led to the formulation of the group ranch concept" [1980c, p. 10].

Many Maasai have presciently used the adjudication process to secure access to a wider resource base by registering kin as members of more than one group ranch, or have ensured private rights to particular areas by arranging for title to individual holdings within group ranch boundaries. Land as property is emerging as the new status base, as invariably occurs during rapid tenurial change [Bohannan 1963]. Acceptance by the Maasai of the group ranch concept has derived from "fear of alternative governmental actions rather than enthusiasm for the proposals" [Goldschmidt 1981b, p. 111]. Group ranch formation has been welcomed as a means of guaranteeing the political integrity of Maasailand by replacement of the spear with title deed [Cossins 1980; Fumagalli 1978; Galaty 1980; Oxby 1982].

To avoid earlier mistakes in the demarcation of grazing schemes, group ranch boundaries were intended to conform to traditional land use divisions of the Maasai. Success in doing so has been erratic, with the boundaries of many ranches matching "no existing Maasai social entity" [Goldschmidt 1981a, p. 54] and nonmembers commonly residing within a group ranch's boundaries [Sandford 1983]. As Cossins observed:

> The original planners of development in Maasailand may well have been aware of. . .the relevance of _enkutot_. . .but as the Masai of today use the term _olokeri_, a small thorn fenced enclosure outside the family kraals to describe a group ranch, it is clear that group ranches do not necessarily enclose intact traditionally functional social or economically viable groups. [1980, p. 2]

Initial years of group and individual ranch formation. Planning the development of a group ranch is the work of Range Planning Teams, which include members from the government's Range Management Division (RMD) and Water Development Department.

> These plans include inventories of natural resources and other assets, such as livestock, dips, boreholes and other improvements. The potential of the rangeland is also assessed and the present carrying capacity (after development) is also estimated. Technical improvements are planned and budgeted as needed. Herd projections are worked out for the ranch herd over a 10-year period, based partly on financial criteria set by the Agricultural Finance Corporation (AFC), and culling and off-take rates are determined. On the basis of this information, stock quotas for the individual members of the ranch are worked out. [Helland 1980c, p. 14]

The first group ranches were formed in the Kaputiei section of Kajiado District, bordering Machakos District. Following the severe drought of the early 1960s, the Maasai living in this area recognized that while nearby, privately operated ranches in Machakos District which had invested in water development had also suffered livestock losses, the numbers lost were slight in contrast to the depletion of their own herds. The residents' desire for developmental changes as found on the private holdings, together with the fact that Kaputiei is one of the more fertile and higher rainfall regions in the district, made it an attractive area for the initiation of group ranches [Hampson 1975]. Following adjudication, 15 group ranches averaging 17,000 ha were incorporated in 1970 and received loans under KLDP I. Table 4.7 shows the sizes and the human and livestock populations of these first group ranches.

Kaputiei was also one of the areas in Maasailand in which privately owned holdings, called individual ranches, were first established. At the time of the grazing schemes during the 1950s, there was a growing demand by prominent Maasai for the subdivision of better-watered areas into individual ranches, and beginning in 1956 several individual holdings of about 800 ha were allocated [Ayuko 1980]. Following the 1960-61 disaster, formation of individual ranches gained momentum and, by 1965, 28 individual ranches comprising about 22,400 ha total had been adjudicated in the Kaputiei area, mostly registered by wealthy or influential Maasai [Goldschmidt 1981b]. By the beginning of the 1970s, altogether about 60 individual ranches, averaging 600 ha and found mainly in this same section of Kajiado District, had received development loans under KLDP I [Wales and Chabari 1979].

Adjudication and registration of both group and individual ranches progressed steadily in Kajiado District during the decade.

> By July, 1981, 65 percent of the district was registered as group or individual ranches (43 group ranches totalling 11,896 sq km and 300 individual ranches covering 1,917 sq km). Excluding the 7 percent of the district allocated to the Amboseli Game Park, the Magadi Concession and agricultural settlement at Ngong and Loitokitok, the remaining area is currently undergoing adjudication, primarily into group ranches. [White and Meadows 1981, p. 4]

Ideally, once a group ranch is in operation, staff of the RMD assist the AFC with post-loan supervision, there is development of water supplies by the Water Department, and

Table 4.7. Area and population characteristics of the first group ranches in Kajiado District

Group Ranch	Area	Number of Ranch Members	Number of Livestock Units	Grazing Density	Livestock per Ranch Member
	-1000 ha-	-heads of households-	--1000---	-ha per livestock unit-	-livestock units per member-
Nkama	39.2	322	7.3	5.4	22.7
Mashuru	30.0	335	4.7	6.4	14.0
Embolioi	23.7	204	2.9	8.2	14.2
Mbuko	18.3	88	3.3	5.5	37.5
Merueshi	18.3	71	1.8	10.2	25.4
Arroi	16.0	113	3.0	5.3	26.5
Kiboko	15.7	67	2.2	7.1	32.8
Empuyankat	15.1	76	2.8	5.4	36.8
Mbilini	14.5	64	2.6	5.6	40.6
Ilmamen	13.1	91	1.6	8.2	17.6
Emarti	12.2	93	1.4	8.7	15.1
Olkarkar	10.2	64	2.0	5.1	31.3
Erankau	9.0	67	1.7	5.3	25.4
Poka	8.9	30	1.8	4.9	60.0
Olkinos	5.9	91	1.5	3.9	16.5
Mean	16.7	118	2.7	6.3	27.8

Source: Bille and Anderson [1980].

disease control measures are implemented though the Department of Veterinary Services. In actuality, development activities have been far from efficiently administered, and intended management controls--limiting of stock numbers, marketing of surplus cattle in rotation, herding of livestock as sex-age aggregates, and so forth--have not taken place.

Group ranches in Narok District. Group ranch formation commenced in Narok District in 1972, and by 1979 two-thirds of the district had been registered, adjudicated, or was under adjudication [Wales and Chabari 1979]. Notable size and land use differences distinguish these group ranches from those of Kajiado District.

> While group ranches are of a respectable size in Kajiado with only 7 with an area less than 10,000 hectares, and 1 with 50 or less members, clearly something went wrong in Narok. Forty-nine ranches have a total area of 1000 hectares or less and of these 34 have areas of 500 hectares or less. There are even 12 "ranches" with an area of 50 hectares or less. . . .In Narok also there are 50 "ranches" with a membership of 50 or less, and of these 40 have 20 or less members with an average of 11 members. [Cossins 1980, p. 16]

Narok group ranches have faced development opportunities and problems not commonly experienced by the group ranches of Kajiado, due to the generally smaller size and cultivable potential of the former. For example, leasing land to wheat and barley producers has become a major source of income for Narok group ranches. However, conflicts have resulted from the leasing of better-watered land which, while not needed by group ranch members in normal years, is required during drought periods [Hjort 1981]. Also, intragroup discord has erupted over the distribution or use of rental income, and long-term costs to the range and to livestock development.

> There are a few instances where the group ranch owners as a whole have decided to hire a contractor, but more often one or a small number of individuals among the owners decide to lease land from the ranch to benefit personally. Disputes among ranch owners concerning whether or not to grow wheat have on occasion prevented the submission of applications for ranch development loans. . . .Yields

> are said to be low and to decline after a few years of cropping, and evidently the risk of crop failure is high in these marginal areas. [Wales and Chabari 1979, p. 19]

Leasing problems are exemplified by circumstances on one Narok group ranch which leased six areas to wheat contractors. By agreement among ranch members, the returns from leasing each of the areas were to go to specified ranch subgroups. This arrangement resulted in inequitable returns, interhousehold quarrels, and a general lack of commitment to livestock development. As Doherty noted:

> Because profits from wheat growing accrue to individuals rather than the group ranch, wheat has not benefited ranch development. Ranch members refuse to make contributions for such development out of their own pockets as the little cash that they do get from wheat is quickly spent. [1979b, p. 35]

Overall, there were two conflicting development alternatives in this instance. On the one hand wealthy members wanted cropping discontinued while, on the other hand, poorer members, with little vested interest in the ranch and nothing to gain from infrastructural improvements for livestock, wanted land leasing expanded. Thus, the more humid setting of much of Narok District, while widening the set of development possibilities physically, has increased the potential for dissension among group ranch members as well.

Change in Tanzania's Maasailand

The course tenurial change has taken in Tanzania's Maasailand is instructive for comparative purposes, given the distinct political and economic philosophies of the two countries. Kenya's economy is relatively open and private enterprise is encouraged, whereas public control of resources and socialistic policies characterize the path taken by Tanzania. A nation-wide program resettling peasants in planned villages has been a hallmark of Tanzania's approach to rural development. Attempts have been made to cast rangeland development efforts also in terms of the nation's political philosophy of Ujamaa.

A plan to establish "ranching associations" in Tanzania's Maasailand was announced in the Range Development Act of 1964, but by 1969 only four of them had been initiated. In that year, the Tanzanian Government and the United States Agency for International Development launched a 10-year Maasai Livestock and Range Management project. Heavy emphasis was given to technical inputs, with the principal objective to increase livestock productivity. The national drive for villagization increasingly overshadowed this Project, until finally with the passing of the National Villages and Ujamaa Villages Act, in 1975, it was decided that ranching associations were unacceptable units for development. Instead, programs to settle Maasai pastoralists into clustered villages, introduce crop production, and establish communally owned herds were begun. The Tanzanian Government enticed the Maasai with housing and permanent water sources, but met with little success, even where the villages were supplied by the government with cattle as communal property [Goldschmidt 1981b]. Pseudo-villages resulted, cosmetic in nature and not leading to any fundamental changes in local organization [Jacobs 1980a]. By the late 1970s, Jacobs found "fairly widespread concern and confusion among both officials and the Maasai as to the desirability and viability of. . .resettlement policies as the most appropriate social innovation to improve pasture control and management of Maasailand. . ." [1978, p. 22]. No meaningful ranching associations were ever fully implemented through which control over the resource base could be exerted. As aptly described by Jacobs:

> Maasai responses to development efforts during this decade were those perhaps best described as a combination of stoic skepticism, thoughtful passive resistance, and traditional pastoral opportunism. Eager to gain land security and access to improved water resources, they displayed immediate interest in ranching associations but were neither surprised nor overly disappointed when the promises associated with them failed to materialize. [1980a, p. 12]

While the Tanzanian Government's approach to pastoral development has been more radical than Kenya's, there is no indication that the Maasai of Tanzania have thereby benefited more or less than their Kenyan brethren.

Summary

Hybrid forms of tenure have been the result of attempts to retain traditional Maasai institutions while meeting national objectives, both political and economic. In Kenya this newly created entity is the group ranch, which is intended to incorporate existing patterns of resource use into a more formal management framework, and thereby promote economic development. A smaller number of individual ranches based on private ownership of land have also been established, but the group ranch is the principal organizational structure by which production is to be expanded. So far, the formation of group ranches has not accomplished this objective, and the primary reason can be found in Fig. 4.1. This flow diagram depicts the principal factors and relationships affecting livestock production by a Maasai group ranch household. As indicated at the extreme left in the figure, "stock quotas, other operational restrictions" are essentially beyond the control of the producer unit. This lack of direct managerial control by producers over resource use is the salient problem of group ranches. This problem, and the potential developmental role of individual ranches on which such control theoretically exists, are examined in the following section.

THE GROUP RANCH PROBLEM

Absence of Sanctions

By the mid-1970s it was widely accepted that group ranches were "much more a concept than a working proposition--and more an exercise in acquiring title to land than an effective means to commercializing beef production" [IBRD 1977, Annex 6, p. 1]. This assessment has remained essentially unaltered to the present [White and Meadows 1981]. Failure cannot be attributed to any single factor, but certainly an incapacity and/or unwillingness to assign liability--by the government to the group ranch, and by the ranch to its members--has restrained development. The absence of required sanctions is evident in the history of ranch development loans and stocking controls.

The impetus for group ranch development is ideally provided through two types of loans, short-term for steer purchases, and long-term for capital investment. Group ranch members are expected to supply 20 percent equity, but they rarely have done so [IBRD 1977]. This failure has not been due to financial inability, for as stated by the World Bank's KLDP II Review Mission, "there is no other community in Kenya with such readily realizable assets which can at a stroke, without loss of benefit to the owner, be used for the benefit of production" [IBRD 1977, Annex 6, p. 18]. Not only have ranches received development funds to which they have not contributed, but the government has had major difficulty with repayment. The land may officially be held as collateral, but realistically foreclosure and selling a ranch because of unpaid loans is a political impossibility. The government's lending agency, the AFC, has therefore had little success in their efforts to coax or coerce individual ranch members to pay back their particular shares of the loans [Helland 1980c].

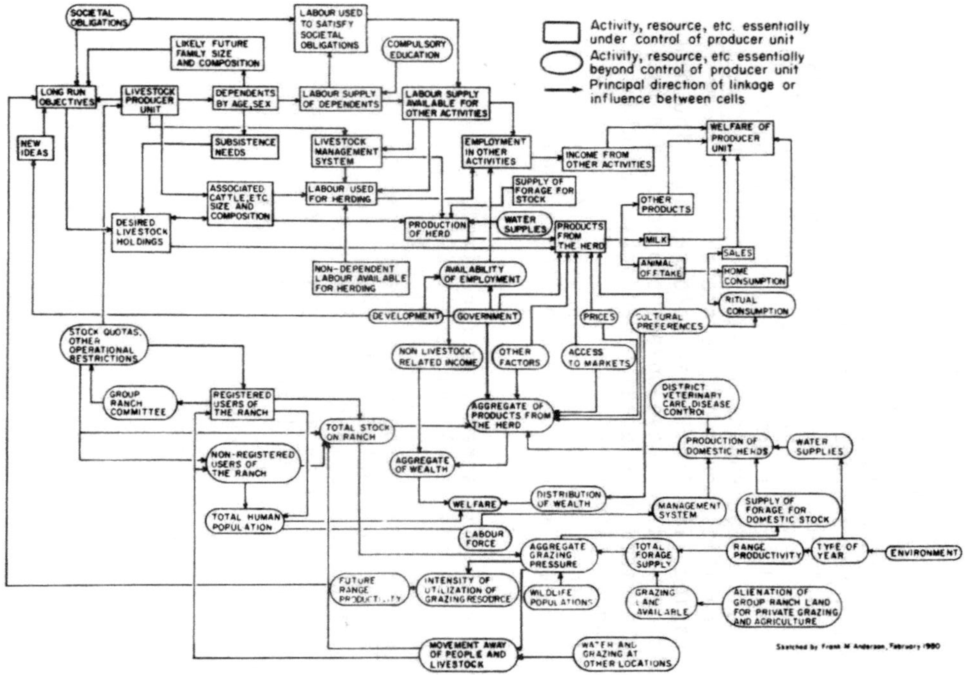

Figure 4.1. Simplified representation of a livestock producer unit in the context of a group ranch

Source: ILCA (1980b).

Fundamentally more serious than widespread failure to contribute to loaned investments and delinquency in loan repayments has been the absence of stock control on group ranches. The allocation of stock rights was originally intended to be included in the adjudication process and enforced by the elected Group Ranch Committee [Davis 1970]. This controversial issue was conveniently left for later consideration, and now the task of livestock unit allocations has been entrusted to an undermanned Registrar of Group Ranches [IBRD 1977]. The overgrazing problem, which hinges on the inability of imposing the group's interest on the individual livestock owners, remains unresolved [Livingstone 1976].

Ownership liability and land use controls were the very issues identified 20 years ago by Jacobs [1963] as key to effective tenurial changes for Maasailand. At the time, he suggested that schemes be incorporated on a cooperative basis. Households of each locality (enkutoto) or group of localities would become shareholders under this plan, leasing land from their corporation. The money derived would be used for development projects, and the land would be divided into grazing paddocks. Eventually shareholders would gain freehold title. Significantly, Jacobs insisted that nonprogressive households not escape the consequences of failure to develop their land:

> One virtue of this scheme, as conceived, is that each individual and each local cooperative will be given an equitable opportunity to make progressive development. And in those anticipated instances where an attempt at development is not made, the individual will be equitably squeezed out, through his legal liabilities (though he will, of course, still be able to sell his shares in the cooperative and, thus, not be left destitute), to make room for more progressive members of the community. For only by some system of legal liability can economic development in any country be realistically pursued. [1963, p. 70, emphasis in original]

The importance of Jacobs' proposal lies in his recognition of control and liability as critical to the development process, an insight verified in the short history of the group ranch. The government is unable to enforce ranch liabilities, and ranches are unable to control their members' stocking practices. Curiously, planners and researchers have often responded defensively regarding this unsatisfactory situation. It is argued that in spite of noncompliance with stock quotas, a measure of development is taking place and needs to be encouraged. This point of view is exemplified by White and Meadows' assessment of group ranch achievements and development priorities:

> Group ranch development has been beneficial to the standard of living of the Masai in Kajiado and development of infrastructure on the newer established ranches should proceed as quickly as possible. Water development should be accorded top priority because better access to water would improve livestock productivity and would create much goodwill among the Masai population. [1981, p. vi].

The need is voiced for "some real Government muscle at the critical points" [IBRD 1977, Annex 6, p. 23], but the call remains largely unheeded. Inaction is justified by a fear of otherwise losing Maasai "cooperation" and the belief that complex socioeconomic change must take time to evolve. Inaction, however, only allows social and economic costs to increase, as population growth aggravates land use pressures and per capita levels of production decline. The assumption that land is less efficiently used when it is not treated as marketable holds true in the case of Maasailand's group ranches [Cohen 1980; King 1977]. They are a short-term palliative, and in their present form can only hinder regional and national economic development in the long run.

Role of the Individual Ranches

The impact of individual ranch adjudication on the rest of Maasailand has been described in both favorable and unfavorable terms. On the one hand, they have been perceived as encouraging class distinctions and the disengagement of individual ranches "from the normal patterns of social and economic reciprocity that have been so vital a factor in the handling of localized disasters" [Goldschmidt 1981b, p. 111]. However, a reduced level of participation in traditional acts of reciprocity can hardly be considered a hindrance to national livestock development. Moreover, the evidence suggests that such disengagement is far from complete. In his study of Maasai households during the drought conditions of the mid-1970s, Campbell [1979b] found a high percentage of individual ranch owners sharing their available grazing with neighboring group ranch members. Informal grazing agreements between sampled group and individual ranches are also described in Chapter V.

More significantly, individual ranches have in a real sense acted as unintended demonstration centers promoting ranch development. This effect is apparent in the growth in interest among group ranch members in the breeding and upgrading of livestock. As Eidheim and Wilson note regarding pastoralists on Elangata Wuas group ranch: "Most of this upgrading among the group ranch members has taken place in recent years, inspired and encouraged by the upgrading achievements made by the individual ranchers" [1979, Annex II, p. 8]. Not all individual ranches are models of progressiveness, and the wide range in management practices which they encompass is exemplified by the sampled ranches (Chapters V and VI). Still, individual ranches have had a positive influence on livestock development in Maasailand.

CONCLUSIONS

Resistance by Maasai pastoralists to organizational and behavioral changes required in the pursuit of national goals illustrates the interaction between economic and political factors in determining social change [Salzman 1981b]. The conflict centers upon tenurial control. The general ineffectiveness of the group ranches in fulfilling livestock development objectives is summarized by Goldschmidt who posits that "at best, the scheme serves to regulate the relationships between the Maasai and the central government, giving them among other things some access to credit and other aids, but it does nothing for the relationships among the Maasai themselves or between the Maasai and their environment" [1981b, p. 112]. Even this rather disparaging assessment is generous, as it is questionable whether a good working relationship exists between ranches and the government. Not only have loan requirements and repayments been largely ignored, but available funds have been left undisbursed out of distrust on the part of group ranch members about the government's intentions. Rather than regard loans for development as a means of increasing ranch productivity "many Maasai describe it as a 'plot' to place them heavily into debt with the result that their land will be sold to 'outsiders'" [Doherty 1979a, p. 6].

With respect to relationships among Maasai, it is not apparent that group ranches have lessened the likelihood of intergroup and intragroup conflict. Strife has mounted between sections as population pressures have increased, exploding at times into violence.[2] Just as volatile, though still simmering, is the unsettled issue of younger men starting families and demanding group ranch membership against the resistence of their fathers and other founding ranch members, given ranch land limits. The problems associated with increasing population pressures have not vanished with the establishment of group ranches. Rather, the formation of the ranches has only postponed their resolution and, concomitantly, delayed livestock development, since mechanisms for decision making at the ranch level have yet to become effective.

Finally, regarding the relationship of the Maasai to their environment, group ranches have been ineffective in altering livestock management practices. As was summarized with respect to Elangata Wuas group ranch:

> The Maasai at Elangata Wuas will continue to be wholly dependent on livestock production with little opportunity for diversification. However, the long-term viability of the production will be uncertain unless controls on the overall stock rate are introduced and enforced. At present, neither the group ranch committee nor the members appear to have a clear understanding of how the ranch should operate, and there is as yet no viable mechanism to control livestock numbers. Given the low and highly variable rainfall in the area, it appears that the ranch is already fully stocked relative to its long-term capacity, so the introduction of limits on further herd growth is crucial. [Bille and Anderson 1980, p. 37]

Traditional Maasai social structure and cultural institutions fundamentally constrain development initiatives [Doherty 1979b; Eidheim and Wilson 1979]. Group ranch formation has not changed this situation, and yet it is the structural framework within which development is expected to proceed. Expanded production will continue to be obstructed unless nominal controls over resource use become genuine.

Meanwhile, the impact of individual ranches on livestock development in Maasailand lends credence to Lewis' [1975] opinion that curiousity, envy, and imitation are essential ingredients for promoting change. One's concept of the feasible depends on experience; aspirations grow with achievement and are influenced by the performance of others [Boulding 1970, 1977; Day 1971; Katona 1964; Simon 1955; Zeleny 1973]. In other words, economies develop by the emulation of the progressive few [Anthony et al. 1979]. However, the positive influence which individual ranches can have on the production practices of group ranch households is limited by the fundamental tenurial distinction. In the following chapter, the livestock management practices of sampled production units are described and qualitatively compared, allowing examination in greater detail of the influence of tenurial and other changes upon the transition to market oriented production.

NOTES

[1]As noted by ILCA scientists, Elangata Wuas is atypical of Maasailand's group ranches both in size and degree of development. It occupies an area of about 77,000 ha, while the average sizes of group ranches developed under KLDP I and II are 17,000 ha and 31,000 ha, respectively. In addition, individual ranches as well as a market center with shops and a dispensary are found within the boundaries of Elangata Wuas. However, these characteristics have not led to pastoral conditions or practices significantly different from those which prevail generally in Maasailand.

[2]The killing of one man and the injuring of several others in a long, ongoing conflict involving three sections in Narok District, was reported in July, 1981:

> The Purko and the Keekonyokie claim they were the original settlers of the Ntulele area and that the smaller Ildamats were immigrants who moved into the area only recently. . . .The Ildamat, however, claim they were the original settlers and now the victims of the expansionism of the two larger groups. . . .At the height of the dispute in 1974, fighting broke out between the Purko and the Keekonyokie on one side and the Ildamat on the other. Seven people were killed and scores were injured before the government could bring the situation under control. [Weekly Review, July 24, 1981]

5 Economic Settings of Producer Samples

The three samples of Maasai livestock producers chosen for study were briefly introduced in Chapter I. These samples and their member units are examined here, with a qualitative assessment of relative commercial orientations as the principal objective. The chapter begins with a prelude on sample selection and data collection. Then, characteristics of the production units considered helpful in understanding and interpreting the production-marketing analyses in Chapter VI are described.

LOCATIONS AND SELECTION OF SAMPLES AND DATA COLLECTION

The Kajiado Group Ranch (KGR) Sample

Selection of the KGR sample and gathering of information were undertaken by the International Livestock Centre for Africa (ILCA), as part of its research of livestock production systems of Subsaharan Africa. The author assisted in data collection, as a member of ILCA's interdiciplinary Kenya Country Programme. Three contiguous group ranches, Olkarkar, Merueshi, and Mbirikani, were chosen for study (Fig. 5.1). The basis of selection was both logistical and analytical, for the region is compact and easily accessible from Nairobi, and yet encompasses ecological and related economic variations of Maasai pastoralism. All three ranches lie within Eco-climatic Zone V but, as indicated in Table 5.1, Mbirikani's setting is more arid than that of Olkarkar or Merueshi. During the survey period, much of Mbirikani suffered a drought, and the rainfall gradient falling southward through the three ranches was in fact steeper than usual [de Leeuw 1982].

Differences in size and population (Table 5.2), and in degree of development, also distinguish Mbirikani from the other two ranches. Olkarkar and Merueshi, registered in 1970, were among the first group ranches established, and received development loans under KLDP I

Table 5.1. Eco-climatic setting of Olkarkar, Merueshi, and Mbirikani group ranches

Eco-climatic Subzone	Annual Rainfall	Potential Evaporation	Proportion of Ranch Area		
			Olkarkar	Merueshi	Mbirikani[a]
	---mm---	----mm-----	-------------percent---------------		
Va	625-750	2,055	95	85	15
Vb	475-625	2,120	5	15	75

Source: de Leeuw [1982], Table 1.

[a]Remaining 10 percent of Mbirikani's area includes patches of Zone VI, and higher-potential lands in the Chyulu Hills and along its southern boundary, where there are swamps fed by water from Mount Kilimanjaro.

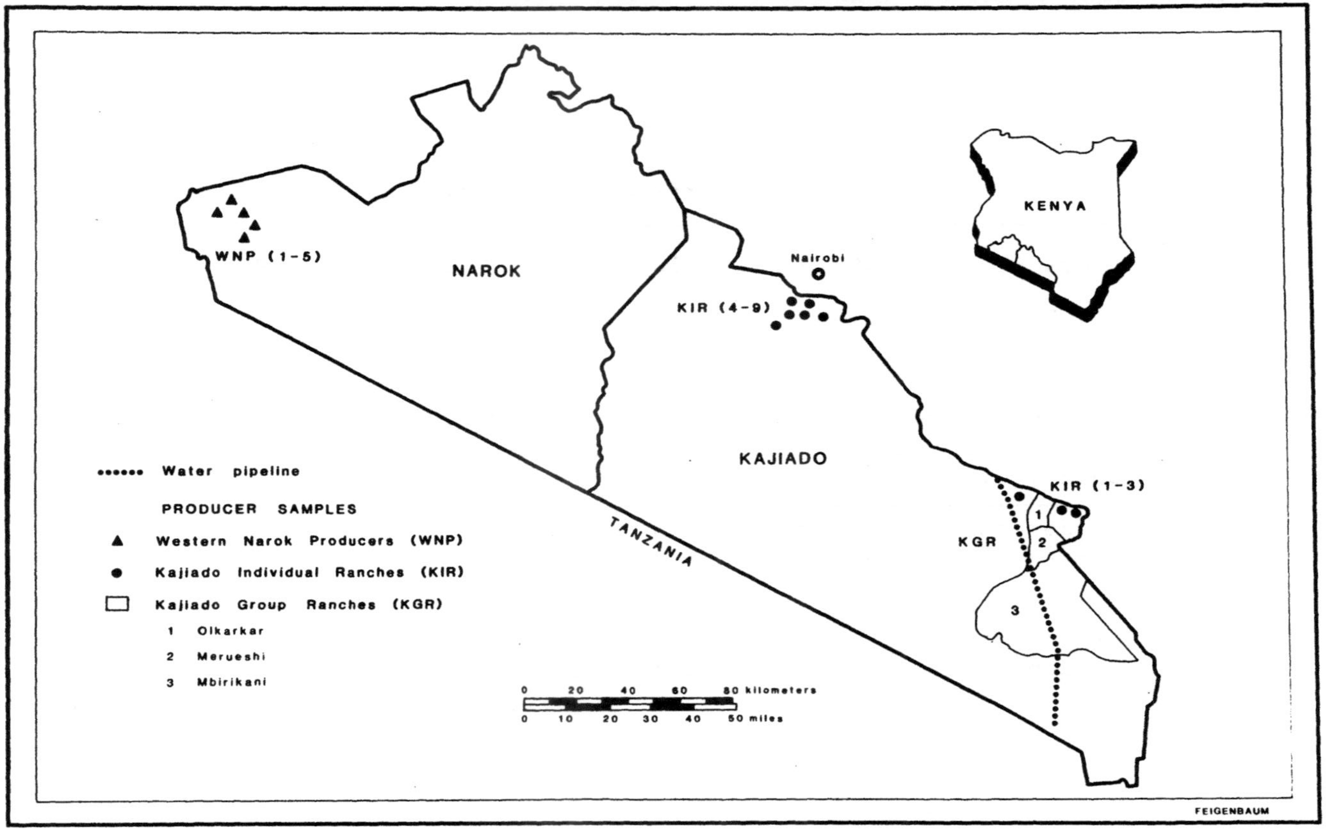

Figure 5.1. Producer samples

Table 5.2. Area, number of households, and livestock populations, Olkarkar, Merueshi, and Mbirikani group ranches, 1981

Ranch	Area	Households	Livestock Population	
			Cattle	Small Stock
	-1000 ha-	--number--	-------head----------	
Olkarkar	10.2	40	3,600	2,300
Merueshi	18.3	36	5,600	3,300
Mbirikani	135.0	206	41,500	19,500

Source: King et al. [1982], Table 2.

for the construction of dips, and on Olkarkar, a water reticulation system. In contrast, Mbirikani group ranch was only registered in 1981, and has yet to receive development loans. Furthermore, the Maasai of Mbirikani belong to the Ilkisongo section, while the Olkarkar and Merueshi residents are Kaputiei-Maasai.

The transhumance of Mbirikani households in contrast to the less mobile production patterns found on Olkarkar and Merueshi reflect distinct eco-climatic differences. On the latter two ranches, Maasai households may live at one enkang for years at a time, while the producers of Mbirikani frequently move as grazing/watering conditions change [Peacock, de Leeuw, and King 1982]. Given their more settled life-styles and longer ranch histories, development interventions would be expected to have had greater impact upon Olkarkar and Merueshi producers than upon Mbirikani producers. A principal objective of ILCA's research has been to test this hypothesis.

An inventory of the three group ranches carried out in 1980 and early 1981 provided the demographic information necessary for household sampling. Surveyed households were ranked by number of livestock per household member, using metabolic-weight equivalents, that is, according to a "livestock equivalent" to "active adult male equivalent" (LE/AAME) ratio.[1] Figure 5.2 depicts the LE/AAME frequency distributions of the households. Three strata of households were delimited according to natural breaks in the distributions: poorer (less than 5 LE/AAME), middle (5 to 12.99 LE/AAME), and wealthier (13 LE/AAME and over), denoted Stratum I, II, and III respectively. Table 5.3 shows the number and distribution of households randomly selected for data collection.

From July, 1981, to July, 1982, data on incomes, expenditures, livestock transactions, livestock management practices, and labor activities of household members were collected on a monthly or fortnightly basis from heads of households and subhouseholds by Maasai enumerators. In addition, milk offtake and calf and small stock weight measurements were periodically recorded for selected herds and flocks. The questionnaires and methods of survey are described in ILCA [1981]. Processing of socioeconomic data has been limited to Olkarkar and Mbirikani group ranches and therefore the sampled Merueshi households are not included in the budgetary analysis carried out in the next chapter. However, discussion in this chapter will include information about Merueshi group ranch and its members for the additional insights into the regional economy thereby gained.

The Kajiado Individual Ranch (KIR) Sample

Approximate locations of the nine individual ranches which comprise the KIR sample, denoted KIR 1 through KIR 9,

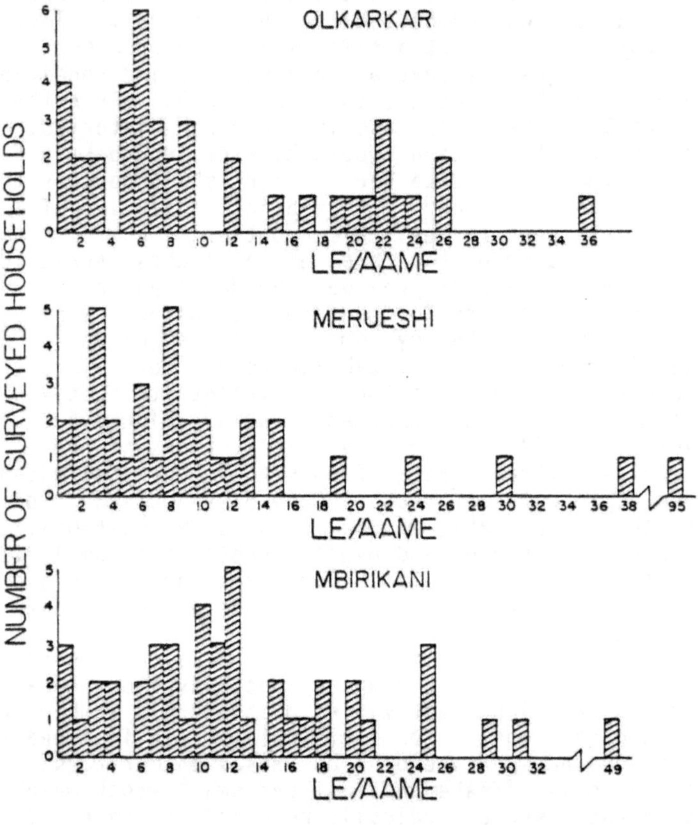

Figure 5.2. LE/AAME frequency distributions of surveyed households, Olkarkar, Merueshi, and Mbirikan group ranches

Source: ILCA [1981], Fig. 3.

Table 5.3. Distribution of Olkarkar, Merueshi, and Mbirikani sample households, by LE/AAME

		Number of Households			
Stratum	LE/AAME	Olkarkar	Merueshi	Mbirikani	Total
I	0-4.99	7	7	6	20
II	5-12.99	9	9	10	28
III	13 and over	8	6	8	22
	Total	24	22	24	70

Source: ILCA [1981], Table 4.

are shown in Fig. 5.1. Three of them, KIR 1, 2, and 3, are in the vicinity of the KGR sample, while the remainder are located 25 to 30 km south of Nairobi. The environment of the latter ranches approaches Zone IV conditions, and thus is less arid area than that of the ranches located near the KGR sample. Information on the operation of these nine ranches was collected by the author through interviews with the owners. The sample size is small and the body of data is neither as extensive nor as comprehensive as was compiled for the KGR sample, but individual ranches representative of widely varying management practices and levels of investment are included.

The Western Narok Producer (WNP) Sample

The WNP sample area was chosen for its distinctly more humid eco-climatic conditions (Figs. 1.1 and 5.1). The hilly terrain of Uasin Gishu Location, Trans Mara Division, in western Narok District, is covered by bushy grassland and broad-leafed trees, and receives 1,000 to 1,200 mm of rainfall annually. As with the KIR sample ranches, information was collected by personal interview from a small number of production units: three holdings which represent traditional, small-scale operations on communal land, and two larger-scale private holdings which are more commercially oriented. Within-sample tenurial differences, in addition to the between-sample ecological differences, permit yet another perspective of factors constraining the transition process in this higher-potential region.

THE KAJIADO GROUP RANCH (KGR) SAMPLE

An appropriate introduction to livestock production and marketing by the KGR sample households is provided by the following review of preliminary findings by ILCA scientists conducting research in the area. Topics discussed, chosen for their relevance to the analysis in Chapter VI, are structure of the livestock populations, water development and resource use, incomes and consumption expenditures, livestock management practices and labor patterns, livestock transactions, and livestock productivities.

Livestock Populations

Mean household size and composition, and mean livestock holdings per household for the three group ranches are presented in Table 5.4. These statistics are based on ILCA's inventory survey of all of the households of Olkarkar and Merueshi and approximately one-half of the households living on Mbirikani. Mbirikani households are on average larger than those of Olkarkar

Table 5.4. Mean household size and composition and livestock holdings per household, Olkarkar, Merueshi, and Mbirikani group ranches

	Olkarkar	Merueshi	Mbirikani
Number of households[a]	40	36	101
Mean household composition			
Household head	1	1	1
Wives	1.6	2.1	2.7
Children, 0-5 years	2.3	2.9	3.3
6-10 years	2.0	2.4	3.0
11-15 years	1.3	0.6	2.8
Dependents, adults	1.9	2.1	1.9
children	0.9	1.8	1.3
Total	11.0	12.9	16.0
Mean livestock holdings per household (head)			
Cattle	98.9	120.7	182.4
Sheep	55.0	61.8	56.3
Goats	43.7	64.2	50.9

Source: Based on ILCA [1981], Table 2.

[a]Includes all Oklarkar and Merueshi households and approximately one-half of Mbirikani households.

and Merueshi, as are the cattle holdings per household. Mean values, however, hide the substantial variability among households' livestock holdings, and given the skewed distribution of holdings, misrepresent the norm. Therefore, in the investigation of herd and flock structures by King et al. [1982], households were stratified using the LE/AAME ratio. Numbers of households and livestock included in this preliminary study are listed in Table 5.5, and Table 5.6 shows the cattle herd structures found, by stratum and ranch. The majority of the stock (two-thirds) are female, with little difference in this proportion among the ranches or wealth strata. A trend toward a higher percentage of steers is evident as one moves from Olkarkar south to Mbirikani, and from poor to rich stratum. A similar prevalence of females, but a smaller proportion of males, is apparent in the flock structures of sheep and goats, shown in Table 5.7.

Ratios of cattle and small stock numbers by ranch and stratum are shown in Table 5.8. Whereas the relative proportions of cattle and small stock are about equal on Olkarkar and for Stratum I households, small stock are held in greater number on Merueshi and by Stratum II households, while the opposite is true on Mbirikani and for Stratum III households. For all three ranches and for Strata II and III, sheep are more numerous than goats. Bulls of improved breeds (mainly local zebu crossed with Sahiwal and, in a few cases, Boran) comprise over one-half of the Olkarkar bulls sampled, a third of the Merueshi bulls, but only 1 percent of those on Mbirikani. However, percentages of improved bulls across wealth strata are more evenly distributed (Table 5.9), suggesting that location has a greater influence on the adoption of breeding improvements than size of livestock holdings. Cattle weights, adjusted for month of weighing and days since the animal last drank, verify the small frame of the Small East African Zebu (SEAZ). SEAZ cows on the three ranches averaged 251 kg and SEAZ steers, 262 kg (Table 5.10). The principal conclusion reached by King et al. is that herd and flock structures on these ranches reflect a management orientation far removed from that one characterizing commercial meat production: "Even the households in the rich stratum, which keep more and bigger steers than the others, have only 23 percent of the herd as castrates of which less than 2 percent are heavier than 300 kg" [1982, p. 27]. While variations among species and sex-age proportions do exist by ranch and wealth stratum, the herd and flock structures fundamentally imply noncommercial objectives.

Water Development and Patterns of Resource Use

Though the infrastructural development of the three ranches has been minimal, water improvements have had

Table 5.5. Distribution of households and number of livestock sampled in the herd structure study, Olkarkar, Merueshi, and Mbirikani group ranches

Ranch and Species	Stratum I	Stratum II	Stratum III	Ranch Total
Olkarkar				
Number of households	2	6[a]	3	11
Cattle	144	328	841	1,313
Sheep	87	365	288	740
Goats	71	284	279	634
Merueshi				
Number of households	3	4	2	9
Cattle	146	300	258	704
Sheep	70	244	209	523
Goats	106	117	151	374
Mbirikani				
Number of households	9[a]	3	9[a]	21
Cattle	300	398	2,418	3,116
Sheep	109	201	1,159	1,469
Goats	122	211	960	1,293
Stratum total				
Number of households	14	13	14	41
Cattle	590	1,026	3,517	5,133
Sheep	266	810	1,656	2,732
Goats	299	612	1,390	2,301

Source: King et al. [1982], Table 4.

[a]Animals from several households inseparable at take-on and herded as one unit.

Table 5.6. Cattle herd structures, by ranch and stratum, Olkarkar, Merueshi, and Mbirikani group ranches

	Ranch			Stratum		
Type	Olkarkar	Merueshi	Mbirikani	I	II	III
	percent					
Female						
Calf	12.2	10.7	8.6	10.7	10.8	9.4
Heifer	19.3	20.5	20.9	18.4	23.5	19.9
Adult, lactating	24.6	24.4	20.4	26.7	22.3	21.2
dry	12.0	10.4	15.9	13.9	12.5	14.6
Total	68.1	66.0	65.8	69.7	69.1	65.1
Male						
Calf (+ young weaner)	9.7	10.8	6.4	8.4	10.4	6.9
Replacement bull	2.8	1.8	2.3	1.9	3.2	2.2
Working bull	2.5	3.1	3.0	3.9	3.1	2.7
Total	15.0	15.7	11.7	14.2	16.7	11.8
Male castrate						
Weaner (+ calf)	8.5	11.6	10.9	11.4	7.0	11.2
Immature	4.6	5.7	10.1	4.2	4.2	10.0
Mature	3.2	0.8	1.5	0.5	2.4	1.8
Large mature	0.6	0.2	0.0	0.0	0.6	0.1
Total	16.9	18.3	22.5	16.1	14.2	23.1

Source: King et al. [1982], Table 12, adjusted for discrepancies.

Table 5.7. Sheep and goat herd structures, by ranch and stratum, Olkarkar, Merueshi, and Mbirikani group ranches

Species, Sex	Incisor Teeth	Ranch: Olkarkar	Ranch: Merueshi	Ranch: Mbirikani	Stratum: I	Stratum: II	Stratum: III
		percent					
Sheep							
Female	Milk	20.8	18.1	22.9	21.4	20.7	21.2
	2-6	31.1	28.3	12.0	24.8	24.6	18.4
	8	16.6	18.8	33.3	21.8	25.4	26.6
Total		68.5	65.2	68.2	68.0	70.7	66.2
Male	Milk	4.7	9.6	8.3	11.7	7.0	7.2
	2-6	2.0	2.7	2.2	2.3	2.7	2.0
	8	0.5	0.4	1.5	1.8	1.3	0.8
Total		7.2	12.7	12.0	15.8	11.0	10.0
Male castrate	Milk	12.7	9.4	9.5	8.3	10.9	10.4
	2-6	8.8	5.4	3.6	4.5	3.7	6.3
	8	2.7	7.3	6.7	3.4	3.7	7.1
Total		24.2	22.1	29.8	16.2	18.3	23.8
Goats							
Female	Milk	16.1	14.7	20.4	21.1	16.2	16.6
	2-6	34.4	35.3	12.5	29.4	23.4	18.2
	8	15.9	15.8	35.5	19.7	26.7	32.5
Total		66.4	65.8	68.4	70.2	66.3	67.3
Male	Milk	6.6	9.6	6.7	8.0	5.6	7.7
	2-6	1.3	1.3	0.7	2.0	0.8	0.8
	8	0.8	0.8	1.0	1.7	1.5	0.4
Total		8.7	11.7	8.4	11.7	7.9	8.9
Male castrate	Milk	8.8	8.3	8.4	7.0	10.8	7.8
	2-6	12.2	11.5	2.9	7.4	8.5	6.0
	8	3.9	2.7	11.9	3.7	6.5	10.0
Total		24.9	22.5	23.2	18.1	25.8	23.8

Source: King et al. [1982], Tables 14 and 16, adjusted for discrepancies.

Table 5.8. Ratios of cattle and small stock, by ranch and stratum, Olkarkar, Merueshi, and Mbirikani group ranches

Ratio (head)	Ranch			
	Olkarkar	Merueshi	Mbirikani	Overall
Cattle:small stock	1.0	0.8	1.1	1.0
Sheep:goats	1.2	1.4	1.1	1.2
Cattle:sheep	1.8	1.4	2.1	1.9
Cattle:goats	2.1	1.9	2.4	2.2
	Stratum			
	I	II	III	Overall
Cattle:small stock	1.0	0.7	1.2	1.0
Sheep:goats	0.9	1.3	1.2	1.2
Cattle:sheep	2.2	1.3	2.1	1.9
Cattle:goats	2.0	1.7	2.5	2.2

Source: King et al. [1982], Table 18.

Table 5.9. Prevalence of improved breeds among working bulls, by ranch and stratum, Olkarkar, Merueshi, and Mbirikani group ranches

Ranch/Stratum	Proportion of Sampled Bulls of Improved Breed
	----------percent-----------
Olkarkar	55
Merueshi	36
Mbirikani	1
I	23
II	23
III	15

Source: King et al. [1982], Table 19.

Table 5.10. Cattle weights, by ranch and stratum, Olkarkar, Merueshi, and Mbirikani group ranches

		Stratum			
Type	Ranch	I	II	III	Ranch Mean
		kg[a]			
Small East African Zebu, adult females	Olkarkar	237 ± 9	237 ± 6	246 ± 8	240 ± 8
	Merueshi	265 ± 7	255 ± 6	261 ± 5	260 ± 5
	Mbirikani	254 ± 5	253 ± 6	252 ± 5	253 ± 5
	Stratum mean	252 ± 4	248 ± 4	253 ± 4	251 ± 4
Small East African Zebu, weaners	Olkarkar	109 ± 20	152 ± 18	138 ± 18	133 ± 18
	Merueshi	160 ± 17	182 ± 14	184 ± 10	176 ± 11
	Mbirikani	204 ± 10	213 ± 11	187 ± 9	202 ± 9
	Stratum mean	158 ± 8	182 ± 8	170 ± 6	170 ± 7
Small East African Zebu, calves	Olkarkar	70 ± 17	72 ± 16	81 ± 16	75 ± 16
	Merueshi	105 ± 14	100 ± 14	94 ± 9	100 ± 8
	Mbirikani	58 ± 9	64 ± 10	62 ± 9	61 ± 9
	Stratum mean	78 ± 7	79 ± 7	79 ± 4	79 ± 5
Small East African Zebu, castrates	Olkarkar	277 ± 47	343 ± 38	314 ± 39	311 ± 39
	Merueshi	181 ± 26	247 ± 27	275 ± 19	235 ± 18
	Mbirikani	243 ± 28	215 ± 35	262 ± 17	240 ± 21
	Stratum mean	233 ± 18	268 ± 18	284 ± 10	262 ± 13
All breeds, adult males	Olkarkar	220 ± 94	285 ± 94	259 ± 90	255 ± 90
	Merueshi	319 ± 67	403 ± 61	406 ± 51	376 ± 47
	Mbirikani	298 ± 49	361 ± 54	348 ± 49	336 ± 47
	Stratum mean	279 ± 37	350 ± 41	338 ± 34	322 ± 34

Source: King et al. [1982], Table 8.

[a]Cattle weights, in kg, are adjusted for month of weighing and days since the animal last drank. Expressed as the mean ± the standard error.

major impact upon settlement and grazing patterns. A pipeline built in 1955 from Loitokitok near Mount Kilimanjaro north to the Nairobi-Mombasa rail line is a principal source of water for Mbirikani and Merueshi households (Fig. 5.1). This pipeline, transecting Mbirikani and passing along the western boundary of Merueshi, was initially equipped with staggered public watering tanks for the Maasai. Since then, private tanks have been added, with livestock owners paying Ksh 1 or Ksh 2 per adult bovine per month for their use, depending whether stock are watered every second day or daily. Calves and small stock are watered freely unless for some reason the producer's adult cattle are not drinking at the tank as well; then, a fee of Ksh .50 per month per small stock watered is charged [ILCA 1981].

In addition to the pipeline, three water tanks have been constructed on the extreme western side of Mbirikani to compensate the Maasai for their loss of access to grazing and watering resources within Amboseli National Park. Merueshi and Mbirikani also contain boreholes managed by the district, one of which on each ranch operated regularly during the survey year. Ranch members are obliged to provide the diesel for the borehole pumps. Finally, on both Merueshi and Mbirikani, water is obtained from hand-dug wells, and on the southern and eastern sides of Mbirikani, from swamps and rivers.

Olkarkar households depend upon a spring at Simba, at the northern end of the ranch, for their water. Using an AFC loan, piping and a pump were installed to bring water to the interior of the ranch, where a tank and dip were constructed but the pump was stolen in the mid-1970s and has yet to be replaced. Simba Springs is the only year-round source of water for Olkarkar, but surface pools during the rainy seasons are utilized as on Mbirikani and Merueshi.

Cattle dips are the other livestock-related improvement on the ranches, with the one used by Olkarkar households located at Simba Springs, and that of Merueshi near the regularly operating borehole. Both were financed by AFC loans. Several dips are found within Mbirikani. Generally, ranch members are responsible for providing their own acaricide or, commonly, they contribute money for its purchase.

It would be remiss in a discussion of resource use to not mention the interranch mobility of producers. Several of the households sampled from Olkarkar and Merueshi were living completely outside their respective ranches (on a neighboring group ranch) at various times during the year. In the case of Mbirikani, extreme movements were recorded, with many of the sampled households off the ranch during the very dry months.

The overriding influence of watering resources upon settlement and grazing patterns is evident in each of the ranches. For example, Olkarkar and Merueshi <u>inkan'gitie</u> are generally found near permanent watering points. Most Merueshi households are settled along the ranch's western boundary near the pipeline, or in the vicinity of boreholes [ILCA 1981]. On Olkarkar, families not living near Simba Springs on the northern side of the ranch must walk up to 10 km to water their animals and fetch water for domestic purposes.

It is on Mbirikani that the impact of water sources upon resource use patterns is most apparent. The Maasai of this more arid ranch have long divided their land into residential and grazing neighborhoods, with the former located near the permanent water sources. However, areas once restricted to use only during rainy seasons are now often grazed at other times, due to the pipeline and tanks. The result has been the disruption of former traditional grazing controls. Peacock, de Leeuw, and King [1982] report efforts by some of the ranch members to more clearly define and enforce ranch interests and rationalize the collective use of the ranch's resources but whether these efforts will be successful is uncertain.

Incomes and Consumption Expenditures

Average nonsubsistence incomes per AAME and income sources for Olkarkar and Mbirikani sampled households are shown in Table 5.11. The dominance of livestock sales on both ranches, but especially Mbirikani, is affirmed. Also, small stock sales on Olkarkar are shown to be significant, with livestock trading a major source of income. Annual expenditures per AAME on food and nonfood consumer items are listed in Tables 5.12 and 5.13, which show food purchases to be by far the most important.

Livestock Management and Labor Inputs

Studies conducted by ILCA included the examination of household labor patterns. Grandin [1982] describes some of the preliminary findings on Olkarkar, involving patterns of labor division within the household and differences in livestock labor inputs by wealth stratum. Age/sex roles in the management of livestock are shown in Table 5.14, and many of the patterns described in Chapter IV are reencountered. Almost all herding is done by children, while adults handle the watering and dipping of stock. In the classification of labor, "boma livestock work" consists of tasks performed in the early morning and evening, such as instructing herders, inspecting and treating animals for disease, and helping kids/lambs find their mothers for suckling and then separating them before grazing or penning for the night. Men do most of the instructing and treating, and some inspecting whereas women and children do the watching and handling of suckling animals [Grandin 1982]. "Other livestock work"

Table 5.11. Mean annual cash income per AAME, Olkarkar and Mbirikani group ranches, July, 1981, to June, 1982

	Olkarkar		Mbirikani	
Source	Amount	Proportion of Total	Amount	Proportion of Total
	--Ksh--	-percent--	--Ksh--	-percent--
Livestock sales	1,112	65	1,523	83
Cattle	997	59	1,474	81
Small stock	115	6	49	2
Livestock trading	319	19	205	11
Remittance	199	12	9	1
Borrowing	41	2	42	2
Milk sales	27[a]	2	0	0
Hides and skins	2	0	9	1
Other	0	0	38	2
	1,700	100	1,826	100

Source: ILCA, Kenya Country Programme. Adjusted for discrepancies.

[a]Equivalent to approximately 10 liters.

Table 5.12. Mean annual food expenditure per AAME, by stratum and per household, Olkarkar and Merueshi group ranches, July, 1981, to August, 1982[a]

	Olkarkar									
	Stratum						Mean of Strata		Mean Household Expenditure	
	I		II		III					
Food Item	Amt	Proportion of Total	Amt	Proportion of Total	Amt	Proportion of Total	Amt	Proportion of Total	Amt	Proportion of Total
	-Ksh-	-percent--	-Ksh-	-percent--	-Ksh-	-percent--	-Ksh-	-percent--	-Ksh-	-percent--
Maize	55	28	108	25	109	24	91	25	840	28
Wheat	1	0	1	0	3	1	2	0	16	1
Sugar	61	31	121	28	130	28	104	29	555	19
Tea	29	15	50	12	38	8	40	11	352	12
Fats/Oils	2	1	12	3	23	5	12	3	124	4
Potatoes	4	2	5	1	11	2	6	2	63	2
Vegetables	1	0	6	2	1	0	1	0	24	1
Meat	3	2	8	2	10	2	7	2	67	2
Other food	3	2	18	4	27	6	16	4	159	5
Subtotal	159	81	329	77	352	76	279	76	2,200	74
Hotel food	11	6	65	15	45	10	42	12	386	13
Hotel drink	26	13	35	8	65	14	41	12	395	13
Total	196	100	429	100	462	100	362	100	2,981	100

Table 5.12. continued

Food Item	Mbirikani									
	Stratum						Mean of Strata		Mean Household Expenditure	
	I		II		III					
	Amt	Proportion of Total	Amt	Proportion of Total	Amt	Proportion of Total	Amt	Proportion of Total	Amt	Proportion of Total
	-Ksh-	-percent--	-Ksh-	-percent--	-Ksh-	-percent--	-Ksh-	-percent--	-Ksh-	-percent--
Maize	211	40	183	54	271	52	222	49	1,771	49
Wheat	33	6	7	2	24	5	20	4	152	4
Sugar	71	14	66	20	60	11	65	14	529	15
Tea	35	6	21	6	35	7	30	7	234	7
Fats/Oils	52	10	17	5	27	5	29	6	231	6
Potatoes	14	3	3	0	7	1	7	2	56	2
Vegetables	7	1	0	0	0	0	2	0	13	0
Meat	20	4	0	0	10	2	6	2	65	2
Other food	7	1	1	0	10	2	6	2	42	1
Subtotal	450	85	298	87	444	85	387	86	3,093	86
Hotel food	46	9	12	4	33	6	28	6	213	6
Hotel drink	30	6	30	9	45	9	36	8	280	8
Total	526	100	340	100	522	100	451	100	3,586	100

Source: ILCA, Kenya Country Programme. Adjusted for discrepancies.

[a]Adjusted to 12 months.

Table 5.13. Mean annual expenditure per AAME on nonfood consumer items, Olkarkar and Mbirikani group ranches, July, 1981, to June, 1982

Item	Olkarkar	Mbirikani
	-----------Ksh-----------	
Tobacco	3	3
Kerosine	6	2
Soap/Detergent	4	5
Transport	13	13
Medical goods and services	12	9
Clothing	34	47
Household items	5	3
Beads	2	2
Total	79	84

Source: ILCA, Kenya Country Programme.

Table 5.14. Age/sex roles in livestock management and domestic activities, Olkarkar group ranch

Age/Sex Group	Observations	Proportion of Task Performed by Age/Sex Group: Livestock Management Activities				
		Herding	Watering	Dipping	Boma Livestock Work[b]	Other Livestock Work[c]
	---number---	---percent---				
Adult, male	245	3	75	68	33	48
female	124	5	6	21	26	16
Child,[a] male	389	47	15	5	18	30
female	363	45	5	5	24	6
Total		100	101[e]	99[e]	101[e]	100
Observations by task		641	123	19	274	64

Table 5.14. continued

Age/Sex Group	Observations	Proportion of Task Performed by Age/Sex Group — Domestic Activities: Milking	Cooking/ Roasting Meat	Childcare	Fetching Wood/Water	Other[d]
	---number--	---percent---				
Adult, male	15	--	6	1	1	1
female	718	80	85	73	89	72
Child,[a] male	35	2	4	2	1	8
female	146	18	5	25	9	19
Total		100	100	101[e]	100	100
Observations by task		137	165	215	143	254

Source: Grandin [1982], Tables 3 and 4.

Note: No differences in task percentages between wet and dry months.

[a]Six years and older.

[b]Includes such tasks as watching small stock and young calves near the boma, instructing herders, inspecting animals for signs of illness, checking for missing animals, treating illness, and helping kids/lambs find their mothers for suckling and then separating them before grazing or penning for the night.

[c]Primarily two activities, looking for lost animals and visiting animals which have been moved temporarily to another area.

[d]Includes cleaning, washing, housebuilding and shopping.

[e]Not 100 percent due to rounding.

includes looking for lost animals and visiting stock which have been moved temporarily to a different area. Domestic tasks, other than roasting meat, are almost entirely performed by women and girls.

The labor devoted to livestock management, by age, sex, and stratum, is summarized in Table 5.15. As shown, the wealthier the household the more total time is spent in livestock management activities, but the less time per livestock unit. Economies of size are not surprising, although as noted by Grandin [1982] the effect is complicated by herd owners' frequent pooling of livestock for grazing, watering, and dipping. The general depiction of labor inputs within Olkarkar households is one in which traditional pastoral roles are predominant. The heavy dependence upon children for herding and the involvement of women in such livestock management tasks as dipping underscore the participation of all family members.

Livestock Transactions

Commercial and noncommercial livestock transactions of sampled households is another area of study for which preliminary findings have been reported. Grandin and Bekure [1982], in their description of livestock acquisitions and offtakes for the Olkarkar and Mbirikani sample households, give particular attention to the significance of market access, the role of traditional expectations in social relations, and the relative importance of traditional and commercial transactions. Transactions are grouped into seven types: purchases, exchanges, and gifts, as forms of acquisition; and sales, exchanges, gifts, and slaughter, as forms of offtake. Reported values of acquisitions and offtakes are producers' estimates. Unintended "emergency" slaughters as when an animal is injured are not included, yet they comprise a significant proportion of all animals slaughtered, 30 percent on Olkarkar and 25 percent on Mbirikani [Grandin and Bekure 1982].

Acquisition and offtake overall. The larger the livestock holdings, the lower the percentage net offtake on both ranches, but particularly on Olkarkar. Noncommercial acquisitions in general and gifts in particular are likely understated by producers. Keeping in mind this source of bias, Olkarkar households of Stratum II, which has the highest number of wage earners, were found to have the highest rate of acquisition. For all strata, purchase was the foremost method of acquisition. The value of acquisitions is greater for Mbirikani households than for Olkarkar households and the types of acquisition are more balanced, with a greater proportion of exchanges and gifts than on Olkarkar.

Grandin and Bekure [1982] summarize the relationship between wealth stratum and the level and type of offtake by noting that while the value of livestock belonging to

Table 5.15. Mean hours devoted to livestock management, by age, sex, and stratum, Olkarkar group ranch

Stratum	Children[a]		Adults		All Workers		
	Male	Female	Male	Female	Proportion of Time	Proportion of Day	Time per Livestock Unit
	---------------hours per day------------------				-percent--	--hours---	-hours per day--
I	5.0[b]	4.6[b]	6.3	0.8	28	3.9	1.29
II	5.7[b]	7.1	6.0	1.5	30	4.2	.60
III	6.4[b]	7.6	7.9	1.5	37	5.2	.25

Source: Grandin [1982], Table 5.

[a]Over the age of six.

[b]Children in these categories spend a mean of approximately two hours a day at school.

the average Stratum III producer on Olkarkar is 10 times that of his average Stratum I neighbor, the former gives away 5 times, exchanges 8 times, and voluntarily slaughters 11.5 times that of the latter. For Mbirikani, the average Stratum III household is 4 times as wealthy as the average Stratum I household, yet total nonsale offtake is only twice that of Stratum I households. Sale is the most important form of offtake on both ranches, but especially on Olkarkar. In particular, there is a notable difference between the two ranches in levels of small stock sales: "Most strikingly, the importance of slaughter in small stock offtake in Olkarkar is less than half that in Mbirikani, while the importance of sale in small stock offtake in Olkarkar is 4 times that of Mbirikani" [Grandin and Bekure 1982, p. 12].

Sales and purchases. Table 5.16 depicts annual sales for the two ranches. Incomes from cattle sales are greater for Mbirikani households than for Olkarkar households, but for small stock the reverse is true. The mean value of individual cattle sold increases, the higher the wealth strata, suggesting poorer households must sell their stock before maturity in order to meet cash needs [Grandin and Bekure 1982]. There is little difference between the ranches or the strata regarding the male/female ratio of cattle sales. In general, three-fourths are males, mostly steers, and one-fourth are females, mostly culled cows. Over the survey year, there was an inverse relationship between rainfall (the availability of milk) and livestock sales, supporting the hypothesis that cash for immediate food purchases is a principal, if not the major, objective of sales [Grandin and Bekure 1982].

Livestock are purchased for building up herds, though the preference among sampled households is to obtain animals for rearing through exchange. As shown in Table 5.17, the mean value of cattle purchased by Olkarkar producers is higher than for Mbirikani producers, since the former purchase mostly males and the latter mostly females. In Table 5.18, percentages of purchases and sales are compared to the percentages of other types of transaction, and the prevalence of commercial transactions on Olkarkar and of exchanges on Mbirikani is apparent.

Livestock Productivity

Cattle. Semenye [1982] presents preliminary results regarding measurements of calf growth and milk production from a study of 362 calves and their dams on the three group ranches. Rates of growth for calves from birth to 120 days averaged one-fourth kg per day. Average total milk yield per cow was found to increase from 3.24 liters at 30 days to 3.62 liters at 120 days, and milk offtake averaged 18 to 25 percent of total yield, or about 0.72 liters per day. Semenye calculated that if the 0.72 liters

Table 5.16. Mean value of livestock sales, by stratum, for sampled Olkarkar and Mbirikani group ranch households

		Cattle			Small Stock		
Ranch, Stratum	Sample Size	Total Sales	Mean Sales per Household	Mean Value per Sale[a]	Total Sales	Mean Sales per Household	Mean Value per Sale
	number of -households-	---head----	--head---	---Ksh----	---head----	--head---	---Ksh-----
Olkarkar I	7	37	5.3	577	29	4.1	120
II	10	95	9.5	874	91	9.1	149
III	7	107	15.2	971	33	4.7	168
Total/Weighted mean	24	239	10.0	816	153	7.1	146
Mbirikani[b] I	8	69	8.6	860	13	1.6	218
II	6	69	11.5	815	16	2.7	188
III	10	149	14.9	927	12	1.2	197
Total/Weighted mean	24	287	12.0	887	41	1.7	202

Source: Based on Grandin and Bekure [1982], Table 4, corrected for errors.

[a]Data adjusted for missing values which represented: Olkarkar: Stratum I, 4.8%; II, 3.8%; and III, 10.8%. Mbirikani: Stratum I, 3.1%; II, 12.5%; and III, 7.5%.

[b]Data for Mbirikani adjusted for 12 months from 11 months of records.

Table 5.17. Mean value of livestock purchases, by stratum, for sampled Olkarkar and Mbirikani group ranch households

Ranch, Stratum	Sample Size	Cattle: Total Purchases	Cattle: Mean Purchase per Household	Cattle: Mean Value per Purchase[a]	Small Stock: Total Purchases	Small Stock: Mean Purchase per Household	Small Stock: Mean Value per Purchase
	number of -households-	--head---	----head-----	---Ksh---	--head---	----head-----	---Ksh-----
Olkarkar I	7	--	--	--	22	3.1	96
II	10	29	2.9	736	23	2.3	100
III	7	10	1.4	925	20	2.9	129
Total/Weighted mean	24	39	1.6	814	65	2.7	107
Mbirikani[b] I	8	25	3.1	771	16	2.0	183
II	6	8	1.3	664	21	3.5	150
III	10	29	2.9	826	30	3.0	130
Total/Weighted mean	24	62	2.6	767	67	2.8	153

Source: Based on Grandin and Bekure [1982], Table 6.

[a]Data adjusted for missing values which represented: Olkarkar: Stratum I, 4.8%; II, 3.8%; and III, 10.8%. Mbirikani: Stratum I, 3.1%; II, 12.5%; and III, 7.5%.

[b]Data for Mbirikani adjusted for 12 months from 11 months of records.

Table 5.18. Summary statistics on mean annual household offtake and acquisition of livestock, Olkarkar and Mbirikani group ranches

Item	Unit	Olkarkar Stratum I	Olkarkar Stratum II	Olkarkar Stratum III	Olkarkar Ranch
Mean value of herd	Ksh	30,530	74,076	261,682	114,547
Proportion of herd value in small stock	%	18	24	11	18
Mean value of total transactions, including slaughter[a]	Ksh	5,989	18,013	30,184	17,694
Proportion of total transactions, including slaughter, in small stock[a]	%	25	23	11	20
Proportion of total transactions which are commercial	%	77	68	63	70
Mean value of commercial transactions	Ksh	4,585	12,338	19,113	11,822
Proportion of commercial transactions in small stock	%	18	13	7	13
Mean value of exchange transactions	Ksh	469	1,430	4,093	1,895
Proportion of exchange transactions in small stock	%	46	40	12	34
Mean value of gifts	Ksh	719	2,938	4,144	2,578
Proportion of gifts in small stock	%	33	41	27	35
Mean value of slaughter[a]	Ksh	215	1,367	2,454	1,314
Proportion of slaughter in small stock[a]	%	100	62	30	65
Mean value of acquisitions	Ksh	579	4,174	4,900	3,182
Proportion of acquisitions in small stock	%	90	17	21	42
Mean value of total offtake	Ksh	5,410	13,839	25,284	14,402
Proportion of total offtake in small stock	%	18	25	11	19

Table 5.18. continued

Item	Unit	Mbirikani Stratum I	Mbirikani Stratum II	Mbirikani Stratum III	Ranch
Mean value of herd	Ksh	47,477	115,669	196,201	129,099
Proportion of herd value in small stock	%	13	19	18	17
Mean value of total transactions, including slaughter[a]	Ksh	21,426	25,421	34,480	29,294
Proportion of total transactions, including slaughter, in small stock[a]	%	17	23	16	19
Proportion of total transactions which are commercial	%	51	53	48	51
Mean value of commercial transactions	Ksh	11,004	13,389	19,391	15,038
Proportion of commercial transactions in small stock	%	7	8	4	6
Mean value of exchange transactions	Ksh	3,995	4,525	9,011	6,058
Proportion of exchange transactions in small stock	%	23	21	18	21
Mean value of gifts	Ksh	4,984	4,585	7,263	5,671
Proportion of gifts in small stock	%	12	28	16	20
Mean value of slaughter[a]	Ksh	1,443	2,923	2,518	2,528
Proportion of slaughter in small stock[a]	%	100	84	96	93
Mean value of acquisitions	Ksh	6,764	5,251	9,261	7,098
Proportion of acquisitions in small stock	%	16	27	14	20
Mean value of total offtake	Ksh	14,662	20,170	29,219	22,196
Proportion of total offtake in small stock	%	18	22	16	19

Source: Grandin and Bekure [1982], Table 11. Adjusted for discrepancies.

[a]Values as reported by producers.

of milk taken for human consumption were fed to calves, their average rate of gain would increase from 0.24 kg per day to 0.56, that is, milk offtake takes place at the expense of increased productivity: "Except for 3 months [during the year] when digestibility of forage available is 65 percent, there is no surplus milk from cows up to their 120th day of lactation for man" [1982, p. 27]. The distance calves must walk for water and the available grazing enroute were found to also critically influence rates of gain. Perhaps most significantly, calf performance across households (as determined by weights at specified ages) varied up to 55 percent of the ranches' averages, reflecting the impact on production of differences in managerial practices.

Small stock. Samples of sheep and goats belonging to 89 households on the three ranches were also studied, to determine linkages between productivity on the one hand and nutrition and behavior on the other [de Leeuw and Peacock 1982]. Like Semenye's [1982] findings for cattle, Peacock [1982] reports a significant variability in production among flocks and individual animals. Using a productivity index defined as the mean ratio of the weight at 150 days of those kids/lambs that reach weaning to the weight of the doe/ewe that delivered, the highest productivity recorded for a Mbirikani household was over 6 times for goats and 2 1/2 times for sheep that of the household with the lowest productivity index, though Peacock cautions against generalizing from the relatively short period of observation. These results support findings reported for Elangata Wuas group ranch discussed in Chapter IV. A marked seasonality in births, with most occurring during the period November-January, immediately after the short rains, suggests a degree of breeding control. However, flock structures indicate a high proportion of castrates beyond optimum selling age, underscoring their importance for home slaughter and noncommercial transactions.

Summary

Characteristics such as predominantly female herd structures, mobile patterns of resource use, low productivity, prevalence of nonmarket livestock transactions, and relatively low levels of market involvement both in production and consumption signify the appropriateness of the KGR sample in representing pastoral livestock production. The general orientation and life-style of the households of Olkarkar, Merueshi, and Mbirikani group ranches reflect objectives and management practices which are widespread in Maasailand. However, clear distinctions by wealth stratum and by ranch have been noted, particularly with regard to offtake rates and frequency of transactions. These differences will be more closely considered in the budgetary analysis in Chapter VI. At this point it is sufficient to note that within the prevailing pastoral mode of production depicted by this sample of Maasai households, distinct variations in producers' activities exist which impinge upon the transition process.

THE KAJIADO INDIVIDUAL RANCH (KIR) SAMPLE

The KIR sample ranches may be described and compared individually with little difficulty, since there are only nine production units. As in the discussion of the KGR sample, the purpose is to present a general depiction of units whereby the analysis in Chapter VI may be meaningfully interpreted. A principal objective is to highlight the variability in management practices and levels of development found within the KIR sample.

Most of the KIR sample ranches were established in the late 1960s and early 1970s, at the same time as the first group ranches (Table 5.19). Their areas range widely, with the largest ranch nine times the size of the smallest. Livestock holdings among the ranches also vary considerably, and small stock numbers especially show uneven distribution. However, an estimation of stocking rates from the data presented in Table 5.19 is not a straightforward exercise, since most of the ranches utilize off-ranch grazing lands. This and other practices are discussed below.

Disease Control and Grazing Practices

Primary management concerns are disease control and securing adequate grazing resources (as is true for the KGR sample households). Diseases often cause heavy losses each year, in spite of preventive and curative treatments. Livestock on several of the ranches are vaccinated annually against such diseases as anthrax and blackquarter, in addition to the compulsory vaccination against foot-and-mouth disease. Other localized diseases are more difficult to control. For instance, tsetse fly infested areas in the vicinity of KIR 2 and 3 make trypanosomiasis a recurrent problem for these two ranches. As another example, KIR 1 and 5 suffer many losses from malignant catarrhal fever every year around March, when the wildebeest are calving. However, for the ranches as a whole the most serious disease problem is East Coast fever (ECF). Cattle losses to ECF and other known or suspected causes of death during 1981 are indicated in Table 5.20. During the same year, small stock losses known or estimated ranged from 20 to over 150 head per ranch. Many small stock are killed by wild animals, such as cheetahs and jackels, although these losses are not as numerous as those due to disease.

All the ranches regularly dip or spray their stock to control ECF and other tick-borne dieases. Small stock, calves, and in some instances improved breeds of adult

Table 5.19. Year established, area, and livestock holdings of KIR sample ranches

Ranch	Year Established	Area	Livestock	
			Cattle	Small Stock
		-ha-	------------head------------	
KIR 1	1966	800	344	320
2	1970	560	412	220
3	1970	680	142	205
4	1977	160	249	1,000
5	1964	1,440	670	2,304
6	1957[a]	238	240[b]	230
7	1969[c]	240	250	640
8	NA	480	350	400
9	1967	720	686	294

[a] Purchased by present owner in 1963.

[b] KIR 6 also has 7 donkeys.

[c] Purchased by present owner in 1975.

Table 5.20. Weaned cattle losses, KIR sample ranches, 1981

Ranch	Deaths		Cause/Suspected Cause
	Number	Rate	
	-head-	-percent-	
KIR 1	63	18	Malignant catarrhal fever; East Coast fever (ECF) and other tick-borne diseases
KIR 2	11[a]	11[a]	Malignant catarrhal fever, ECF, and accidents
KIR 3	16	11	ECF and predators (lions)
KIR 4	20	12	ECF
KIR 5	41	6	Malignant catarrhal fever and ECF
KIR 6	5	4	ECF
KIR 7	42	17	ECF
KIR 8	"Many"	--	ECF
KIR 9	20	3	ECF

[a]Owner's herd only.

cattle are given anthelmintics to combat internal parasites. These treatments are summarized in Table 5.21. In addition, general antibiotics are commonly administered on the ranches to animals which appear ill.

Table 5.22 shows the extent of off-ranch grazing by the KIR sample ranches during the survey year. Livestock belonging to eight of the nine ranches grazed outside ranch boundaries during 1981. Only two of the ranchers do not normally make use of outside grazing or avail their land to others: the owner of KIR 8, the perimeter of which is fenced, and the owner of KIR 6, who believes outsiders "steal" his grazing anyways. Informal reciprocal arrangements are common. KIR 1, the only ranch in the sample which was grazed by outside livestock during 1981, carried small stock belonging to neighboring ranches during the earlier part of the year when the forage cover was abundant. Six months later when no grass was left on KIR 1 or the neighboring ranches, the owner or KIR 1 had to pay to have his livestock grazed on land 50 km away. Reciprocal or contractual grazing arrangements by the ranches may serve to relieve the constraints placed upon production by the spatial variability of the resource base, but as KIR 1's circumstances illustrate, such agreements entail risks.

While outside grazing is common, supplemental feeding is used to a very limited extent on the ranches. The owner of KIR 5 sometimes feeds branmeal or maizemeal for a few days to weak animals recovering from disease, and the owner of KIR 8 at times gives maize stalks, cut from fields 10 km away for Ksh 35 per acre, to the ranch's livestock. But supplemental feeding is not practiced on any of the ranches on a regular basis, and on most not at all.

Investment and Orientation

Disease control measures and reliance upon off-ranch grazing resources are practices common among the ranches. However, the nine ranches span a wide development spectrum, as reflected in level of capital investment and management orientation. At one extreme is KIR 9, where substantial investments have been made in a dip, reticulation system, fencing, weighbridge, and other capital improvements, and the perspective is entirely progressive. At the other extreme one finds KIR 2 and 3, with thorn bush fenced livestock enclosures essentially the only capital investments. KIR 2 may have been more progressively managed at one time, for in 1973 one-half of the perimeter was fenced. However, the fence has since fallen into disrepair, and today the level of development on the two ranches is for all purposes indistinguishable from that found on group ranches.

The recent history of a dip built jointly in 1970 for KIR 2 and 3, plus a third nearby ranch, indicates the prevailing mood of resignation and absence of initiative

Table 5.21. Frequency of use of acaricides and anthelmintics, KIR sample ranches, 1981

Ranch	Use of Acaricides	Use of Anthelmintics
KIR 1	Cattle dipped weekly; small stock dipped every other week.	Small stock and calves drenched twice a year.
KIR 2	Cattle dipped weekly; small stock sprayed every other week.	Small stock and calves drenched twice a year.
KIR 3	Cattle and small stock dipped weekly.	Small stock and calves drenched twice a year.
KIR 4	Cattle dipped weekly; small stock dipped every other week.	Small stock and calves drenched four times a year.
KIR 5	Cattle dipped weekly, twice weekly when rains begin; small stock dipped every other week.	Small stock, calves, weak cattle, purchased steers drenched four times a year.
KIR 6	Cattle sprayed weekly; small stock sprayed every other week.[a]	Small stock, calves, weak cows drenched twice a year.
KIR 7	Cattle dipped weekly; small stock dipped every other week.	Small stock and calves drenched four times a year.
KIR 8	Cattle dipped weekly; small stock dipped every other week.	Small stock and calves drenched three times a year.
KIR 9	Cattle and small stock weekly (twice weekly during rains or ECF outbreak).	Small stock, calves, 20 high-grade cattle drenched three times a year.

[a]Dip constructed in early 1982.

Table 5.22. Off-ranch grazing, KIR sample ranches, 1981

Ranch	Period and Circumstances	Cost or Form of Agreement
KIR 1	Three months, Aug to Oct, all of the cattle except 40 to 50 of the cows, on land near the Chyulu Hills. Dip belonging to a nearby cooperative ranch was used.	Ksh 5 per head per month
KIR 2	Four months, Aug to Nov, all livestock on neighboring group ranch land.	Reciprocal[a]
KIR 3	Four months, Aug to Nov, all livestock on neighboring group ranch land.	Reciprocal
KIR 4	Full year, 400 ha rented from nearby cooperative.	Ksh 5000 for the year
KIR 5	Full year, 260 ha,[b] neighbor's land.	Reciprocal
KIR 6	Half year, state land.[c]	N.A.
KIR 7	Half year, neighbor's land.	Reciprocal
KIR 8	No outside grazing.	--
KIR 9	Full year, 480 ha,[d] neighbor's land.	Reciprocal

[a]Reciprocal agreements may simply involve the sharing of available forage, as with KIR 2, KIR 3, and KIR 7, or can include arrangements for the use of watering and dipping facilities as well, as in the case of KIR 5 and KIR 9.

[b]Increased to 320 ha in 1982.

[c]This rancher usually practices deferred grazing on his ranch, but because of illness, management during 1981 was lax.

[d]From three individual ranches, 120 ha, 280 ha, and 80 ha. On this land were grazed 340 AFC-financed and 70 homebred steers.

to expand production. The dip has not been in operation since 1980, when parts were damaged and stolen. Yet, the three ranchers pay to use dips which belong to neighboring group ranches rather than arrange for their dip's repair. In addition to the financial costs entailed by the use of the group ranch dips, the ranchers recognize that the productivity of their stock has declined due to the longer distances that they must walk. Meanwhile the loan which financed their own dip's construction has still to be repaid.

Investments in fencing on the ranches also illustrate the range in conditions and attitudes governing capital development. The owner of KIR 9 is at the forefront of positive change. The fencing of the perimeter of this ranch, completed in 1979, cost Ksh 180,000, and Ksh 120,000 per year is required for its maintenance. The owner of KIR 9 believes that the sizable expenditures are justified. In his opinion, fencing has contributed significantly to the productiveness of the ranch, and is an essential step to the ranch's further development. In support of this view, he cites a sharp decline in livestock losses: 20 to 40 percent of the livestock were lost each year to malignant catarrhal fever before the fence was constructed, since the area was evidently a favorite calving ground for wildebeest, but not a single animal has died from this disease since the fencing of the ranch's perimeter. In addition, the fence has prevented grazing competition by game, and greatly aided tick control. The next stage in the ranch's development will be the paddocking of pastures over the next three years, which will permit pastures to be grazed more efficiently as well as preclude the need for corralling livestock at night. The owner of KIR 9 intends to reduce the cost of paddocking by using electrified fencing which require fewer strands, another example of his innovative orientation.

The only other ranch in the sample with a fenced perimeter is KIR 8. The owner of this ranch likewise believes that reduced disease losses and an enlarged grazing resource (by the prevention of access by game and outside livestock) have resulted in a net gain. But, benefits and costs have not been even roughly quantified, a surprising situation since fencing is perhaps the largest investment decision for a rancher.

Depending upon the particular circumstances, the perceived benefits may not always justify the expenses involved. As related in the case of KIR 2, costs exceeded expected gains on this more traditionally oriented ranch, and initial fencing was not completed or maintained. As another example, for the perimeter of KIR 5 to be fenced in 1981 would have cost approximately Ksh 400,000, and in the opinion of the owner, would not have been worth it: "The fence does not produce anything." Construction and expected maintenance costs of the fence over a specified time horizon would need to be compared with the expected benefits gained from greater disease and grazing controls for this conclusion of negative net returns to be verified. While the perimeters of only two of the ranches have been fenced, the use of purchased materials for the fencing of night bomas and grazing enclosures is common. A small paddock was fenced in 1980 on KIR 1, to assure ample grazing for young stock; the owner of KIR 5 constructed an enclosure using cyclone fencing; and on KIR 6 one for purchased steers was in the process of being constructed, using poles and rails.

Breeding stock expenditures also vary widely among the ranches, though not as dramatically as differences in fencing investments. Sahiwal and Sahiwal-crossed bulls are most frequently used to upgrade cattle herds (Table 5.23). However, on some of the ranches other breeds have been introduced, with KIR 9 again leading in innovations. Simmental bulls, which cost Ksh 10,000 each and were chosen for the breed's good rate of gain, a Dorper ram imported from Botswana for Ksh 8,000, and the introduction of Toggenburg goats are among the breeding innovations found on KIR 9. The owner of KIR 1 intends to develop a dairy herd by using artificial insemination to produce Friesian crosses, and the owner of KIR 4 is also planning a dairy operation, utilizing bulls of unspecified European breed. The owner of KIR 7, on the other hand, is concentrating on producing slaughter stock from Boran bulls.

Differences in levels of investment and management orientations are matched by a broad range in bookkeeping practices. Records on ranch operations run from nonexistent to precise and up-to-date. For example, the owner of KIR 1 keeps no written records of how much is spent on medicines for his livestock, a situation representative of all of the ranches with low levels of investment. There is a conspicuous absence of basic record keeping even for some of the more developed ranches. The owner of KIR 8 could only grossly estimate the ranch's livestock numbers, and took the casual attitude that "if some get lost, they are branded." Moreover, KIR 8's small stock sales are not recorded, and though many livestock succumb to East Coast fever and other diseases, no records are kept of livestock deaths: "It rains, animals die."

In contrast, all expenditures and incomes for KIR 9 are carefully monitored by the owner, to the fraction of a shilling that it costs to dip or drench one animal. Production parameters such as calving rates and weight gains are recorded, and all slaughter stock are sold by liveweight using the ranch's weighbridge. The distinct differences in bookkeeping practices, more than any other qualitative measure, signify the range in management of these individual ranches.

Table 5.23. Breeds of improved sires, KIR sample ranches

Ranch	Bulls	Rams/Bucks
KIR 1	Sahiwal Sahiwal-zebu[a]	
KIR 2	Sahiwal Sahiwal-zebu	
KIR 3	Sahiwal-zebu Boran-zebu	
KIR 4	Sahiwal-European	
KIR 5	Friesian Ayrshire Boran	Dorper
KIR 6	Sahiwal-zebu	
KIR 7	Boran	
KIR 8	Sahiwal-zebu	
KIR 9	Simmental	Dorper Toggenburg

[a]Zebu refers to Small East African Zebu.

Steer Fattening and Livestock-Derived Incomes

Despite differences in degree of development and conscientiousness of management, most of the ranches operate similarly in the regular purchase and sale of AFC-financed steers. Livestock purchases and sales for the ranches during 1981 are indicated in Table 5.24, and in Table 5.25 their AFC-financed steer grazing operations are shown. The loans may be used to purchase steers locally in Maasailand, or from holding grounds to the north, where stock purchased and quarantined by LMD, as described in Chapter III, are bought on a liveweight basis. Some ranchers prefer local purchases, expecting fewer disease problems, while others find buying LMD steers--a single purchase by weight of animals of one age group--a more attractive arrangement.

In 1981, the LMD cattle sold for about Ksh 3.40 per kg, and the average number of AFC-financed steers purchased at one time by the ranches was 100 head. All of the herds indicated in Table 5.25 were obtained from the LMD's Kirimun Holding Ground in Samburu District (except those of KIR 9 which were purchased at another LMD holding ground), and rail transport costs to the KIR sample ranches averaged Ksh 100 per head. AFC steer loans in 1981 were made at an interest rate of 10 percent (in 1982 the rate increased to 13 percent), with payments due twice a year. However, rescheduling of loans is frequent if a large number of the steers are lost such as to disease. In general, grazing older steers over a short period of time is preferred (Table 5.26). Finishing steers on grass is a rational strategy, given the low prices for feeder cattle relative to feed prices [Meyn 1978; Schaefer-Kehnert 1979]. As the owner of KIR 5 notes, grazing steers is all the more attractive when the risk of drought is considered: "One can sell steers, but it is difficult to sell a cow with a calf."

The importance of stock age and the timing of sales to a successful steer-fattening strategy is illustrated by KIR 9's transactions for the survey year. Prior to 1981 the owner of KIR 9 had always grazed older steers for a short period. For the first time, he decided to feed younger steers over a longer span of time. The steers were purchased in December, 1980, for Ksh 1,200 each and they weighed on average 210 kg. They were sold to other ranches for further fattening in February, 1982, for Ksh 1,800 each, but only because the owner of KIR 9 could not afford to carry them further into the dry season. The steers lost an average of 40 kg during their final three months on the ranch, dropping from 330 kg to 290 kg. The returns foregone by not selling earlier are examined in Chapter VI. Following this experience, the owner of KIR 9 vowed to return to a strategy of purchasing only older

Table 5.24. Livestock purchases and sales, KIR sample ranches, 1981

Ranch	Purchases	Sales
KIR 1	200 steers[a]	100 steers[a]
KIR 2	None	10 cows, 10 steers
KIR 3	100 steers,[a] 7 small stock	110 steers[b]
KIR 4	100 steers[a]	10 cows, 40 steers[a]
KIR 5	149 steers, 10 small stock	70 cows, 160 steers, 357 small stock
KIR 6	None	20 cows
KIR 7	100 steers[a]	2 cows, 85 steers,[c] 130 small stock
KIR 8	None	5 cows, 6 steers[d]
KIR 9	340 steers[e]	50 cows, 115 small stock

[a]AFC-financed.

[b]Includes 100 AFC-financed, purchased in 1980.

[c]Includes 70 AFC-financed.

[d]Small stock were sold, but number could not be estimated by owner.

[e]AFC-financed, purchased in December, 1980.

Table 5.25. Use of AFC steer loans, KIR sample ranches, 1981

Ranch	Purchase			Sale		
	Date	Number	Price	Date	Number	Price
		-head-	-Ksh-		-head-	-Ksh-
KIR 1	March, 1981	100	900	July, 1981	100[a]	1,600
	June, 1981	100[b]	900	--	--	--
KIR 2[c]	--	--	--	--	--	--
KIR 3	Spring, 1981	100[d]	800	--	--	--
KIR 4	March, 1981	100	800	Jan, 1982	40[e]	1,800
KIR 5[f]	--	--	--	--	--	--
KIR 6[g]	--	--	--	--	--	--
KIR 7	March, 1981	100	800	Nov, 1981	70	1,600
KIR 8	--	--	--	--	--	--
KIR 9	Dec, 1980	340[h]	1,200	Feb, 1982	330[h]	1,800

[a]Sold to a trader exporting live animals to the Middle East. Owner received an exceptionally high price. Similar steers on nearby group ranches were selling for about Ksh 1,100 each at the time.

[b]Owner intended to sell these steers in Feb, 1982, to the Kenya Meat Commission.

[c]Owner last used an AFC loan to purchase steers in 1973; feels ranch is presently well stocked.

[d]Twelve steers died. Owner planned to sell the remaining 88 in early 1982.

[e]Twenty of the steers died. The forty were sold in Jan, 1982, to pay off part of the loan and to buy younger animals. Owner planned to sell remaining 40 after the rainy season, in order to profit from expected weight gains.

[f]Does not use AFC steer loans.

[g]Owner last used an AFC loan to purchase steers in 1977; 73 head were bought locally for Ksh 250 each and sold one year later for Ksh 750 each. Now that the ranch has a newly constructed dip, the owner intends to purchase steers using an AFC loan in 1982.

[h]Transactions not included in the ranch budget in Chapter VI.

Table 5.26. Preferred length of period for feeding steers, KIR sample ranches, 1981

Ranch	Preferred Period
KIR 1	Six months
KIR 2	--
KIR 3	Six months
KIR 4	About one year
KIR 5	Three to eight months
KIR 6	Six months
KIR 7	Six to eight months
KIR 8	Six months
KIR 9	Nine months or less

steers, and grazing them over a period of nine months at most.

The incomes for all the ranches derive mainly from livestock sales, principally steers. On most of the ranches, cows are milked only for home consumption, but KIR 4, 6, and 8 each sell between 10 and 20 liters of milk per day at the nearby town of Kiserian. Also, ranches KIR 4 through KIR 8 sell manure, some of them more regularly than others. Farmers, especially coffee growers, purchase the manure, with prices ranging from Ksh 400 to Ksh 500 per pickup load. For example, from one night enclosure the owner of KIR 4 sells three loads per month, and two loads per month are sold by the women of KIR 6. The owner of KIR 5 prefers to spread the manure on his land, as does the owner of KIR 9, but a relative in need of money will often be permitted to sell whatever supply there is at the time.

Summary

Clearly, the KIR sample ranches which are favored ecologically and are stronger financially have a developmental advantage over the others. But more importantly, the ranches which are making a successful transition to commercialized production are those having a progressive management orientation. Private tenure, per se, is apparently not a sufficient condition. Rather, innovative management, made possible by the explicit control over production allowed by private tenure, is the crucial criterion. KIR 9's operations, in particular, represent development possibilities when livestock management practices widen rather than constrain production-marketing opportunities. The development plans of some of the other ranches exemplify this progressive orientation as well. For example, the owner of KIR 5 is considering leasing a large area of his ranch to a contract wheat farming operation, which would provide income for infrastructural development of the rest of the ranch. Moreover, lands would be fenced and following a leasing period of perhaps six years could be converted into improved pasture. Concurrently, the relative undeveloped ranches included in the sample are being managed with an outlook well characterized by the owner of KIR 1: "You don't die, but never really get ahead either." In sum, examination in Chapter VI of differences in productivity, investment, and marketing levels within the KIR sample, itself, may be as revealing of production constraints as comparisons with the KGR and WNP samples.

THE WESTERN NAROK PRODUCER (WNP) SAMPLE

Of the five production units which comprise the WNP sample, three are traditional household operations located on trust land or group ranch land. These households, denoted WNP 1, 2, and 3, are representative of most producers in the area. The other two production units, designated WNP 4 and 5, are larger, recently acquired and privately owned. Their operation represents the production possibilities of the area, given freehold tenure, funds for ranch development, and commercially oriented management. Since WNP 4 and 5 are only in the beginning stages of development, neither of these ranches can be evaluated as established, ongoing enterprises like the KIR sample ranches. Still, the analysis of their livestock production and marketing activities and development strategies affords insight into perceived development options, given the area's nascent state of commercial ranching.

General Features

Characteristics prevalent for the sample area as a whole are noted, before discussing each of these production units individually. First, traditional livestock management objectives and practices differ little from those of the KGR sample households. Livestock accumulation is the goal, with cattle kept primarily for milk production and long-term security. Breeding is unrestricted, and animals are herded by day, watered at streams, and corralled at night within thorn bush fenced enclosures. The enclosures and the spray runs used for applying acaricides are often the only capital livestock investments. A typical cattle enclosure may take a week to ten days to build, is repaired as the need arises, and can be expected to last about four years. Some producers such as the head of WNP 2 are now using wire fencing.

The impact of the higher-potential environment on livestock management practices is apparent. Securing adequate grazing and watering resources is not the major concern in the WNP sample area, unlike much of the rest of Maasailand (although reports of bush encroachment over the past 20 years indicate instances of overgrazing). Rather, principal concern is with disease problems, especially East Coast fever and other tick-borne diseases, trypanosomiasis, and internal parasites. Spraying of livestock to control ticks is more common than dipping, with spray runs built using local materials.

The bushy terrain of the WNP sample area provides an ideal environment for the tsetse fly, and trypanosomiasis is a major though localized problem. For example, WNP 1 and 2 are located in areas where fly infestation is not heavy, whereas the livestock of WNP 3, 4, and 5 suffer frequently and sometimes severely from the disease. Internal parasites are treated by the regular drenching of animals. The head of WNP 2, for example, observed a definite improvement in the health of his cattle after treating them for the first time during the survey period, at a cost of about Ksh 3 per animal. Thereafter, he intended to drench his stock regularly every four months. As elsewhere in Maasailand, antibiotics are widely administered, and the Department of Veterinary Services conducts compulsory foot-and-mouth vaccination campaigns twice a year. Also to be mentioned is a lively trade in black market medicines, especially acaricides, with drugs often costing one-half the official price. However, legal penalities are not the only risk, since the drugs are sometimes other than claimed and ineffective or even harmful to animals.

The more humid setting of the WNP sample area ease watering and grazing problems and tend to accentuate disease problems. But the major eco-climatic impact is in the cultivable potential of the area. Livestock production is dominant but the cultivation of crops is transforming the WNP sample area from a pastoral region into one of mixed farming. Factors facilitating this transformation, in addition to the favorable environment, are the sedentary life-style of the Maasai of the area, their willingness to practice cultivation, and the ease with which agricultural labor is obtained from neighboring non-Maasai districts.

The Maasai of the region live relatively settled lives, with movement occurring even less frequently than for the Olkarkar households of the KGR sample. Permanent or semi-permanent residence, made possible by sufficient and adequately distributed grazing and watering resources, has permitted crop cultivation, which in turn has enhanced the settled life-style. Cultivation most commonly is by oxen-pulled plough. Intercropping, especially of maize and beans, is common, though some producers claim single crop stands result in higher yields. None of the sampled producers could remember a crop failure due to drought, but all mentioned years of reduced yields when the rainy season was late or lighter than usual. The area is abundant with wildlife, and game (and livestock when left unattended) pose the principal threat to crops. For example, the head of WNP 2 did not plant beans in the budgeted crop year because the previous crop had been eaten/destroyed, mainly by bushbuck. Likewise, previous losses dissuaded the owner of WNP 4 from planting any crops during 1981.

Losses to wildlife reflect the area's relatively unsettled state and low population density. If the land were occupied by farmers, as is the case in neighboring Kisii and South Nyanza Districts, there is little doubt that the WNP sample area would be extensively cultivated. As it is, district (tribal) boundaries have prevented the expansion of cultivators into the area from these densely populated districts, and the contrast in land use across district lines is readily apparent. Within the same eco-climatic zone, land on the non-Maasai side of district boundaries is a patchwork of smallholdings, while wooded pastureland covers the Narok District side, disrupted only by an occasional field of maize and beans near a homestead.

Neighboring populations provide a labor pool from which the Maasai liberally draw, filling labor shortages as well as tempering whatever cultural aversions exist to the cultivation of the land. As exemplified by the three small-scale operations sampled, farming labor is contracted on a tenant labor basis (despite the land not being privately owned). Men, usually with their families, contract to work for Maasai households, dividing their labor between cultivating crops for the Maasai and for themselves and their families. Short-term labor is not uncommon for specific tasks, such as weeding, clearing and fencing, but longer-term "resident" labor is the norm.

Exploiting the area's cultivable potential and nearby surplus labor markets, the Maasai are expanding beyond subsistence maize production to cash crops, including beans, cabbage, onions, sugarcane, and pyrethrum. Households which previously had only grown maize and beans for home consumption are now producing for markets on a regular basis. Crops which are marketed are most frequently sold at the homestead to local consumers or traders, or in nearby towns to parastatals such as the National Cereals and Produce Board. Sales of maize and beans are by 90 kg bags, while local sales are usually by the _debe_, a 20 liter container. Six _debes_ are considered to be the equivalent of one bag. Better prices are often obtained locally, but selling to the marketing board is attractive to producers because all of a crop can be sold at one time.

The tenant labor system operates with the understanding that the non-Maasai populations will not be allowed to settle in the area on a permanent basis. The Maasai therefore feel no threat of land alienation. Moreover, tribal barriers to non-Maasai settlement are unlikely to recede in the near term even with increased integration of the Maasai into the national economy, since ethnocentric and nationalistic identifications of the Maasai are not negatively correlated [Laughlin 1980].

Some group ranches have been formed, such as in the vicinity of WNP 3, but the tenurial future for much of the sample area is still undetermined. Local land adjudication committees unofficially allocated user rights among residents in the late 1960s, based upon settlement patterns at the time. Formal boundaries were expected to be drawn soon thereafter. However, central government officials have preferred to leave the demarcation issue to local decision, and since the Maasai have felt secure in their rights to future title, there has been little impetus locally for adjudication. Still, in recent years there has grown a greater desire for formal deeding of land. Producers in the vicinity of WNP 1 and 2, for example, think it likely that the land will be demarcated into individual holdings within the coming five years, though opinions are expressed cautiously. Significantly, the formation of group ranches is neither expected nor desired, a situation attributable to the expansion of cropping enterprises and recognition of investment opportunities that will accompany individual ownership.

When asked how freehold tenure would affect his livestock holdings, the head of WNP 2 admitted that he would need to reduce the size of his cattle herd, from 160 animals to about 60, since much of the communally grazed land presently used would no longer be available. Yet this producer looked forward to closure, intending breeding and pasture improvements once he held title to his own land. This attitude is not unusual in the area. As households have become increasingly market oriented, private ownership of resources has gained favor, with fencing of land and upgrading of herds common forms of investment envisioned. Given this background on the sample area, characteristics of each of the production units are now briefly discussed.

WNP 1

This household is headed by a young man of about 30 years. He supports eight dependents: two wives, four young children ranging from 6 months to 3 years (two by each wife), and two older girls, 8 and 12 years, who are relatives of one of the wives. The household is situated on trust land, and was established in 1976 when the head of WNP 1 married his first wife and left his father's home. A brother and his family live close by, and though they maintain separate households, the brothers' stock often graze together. Also, with his brother to shoulder many of the livestock management responsibilities, the head of WNP 1 has been able to concentrate upon the buying and selling of cattle and the growing and marketing of cash crops. Over 110 head of cattle were purchased and sold during the first 10 1/2 months of 1981, and crops of maize, beans, onions, and cabbage were marketed, the onions in Nairobi. These trading and cropping enterprises are examined in Chapter VIII.

The head of WNP 1 represents small-scale producers of the area who are using markets for cattle and crops in order to earn money to expand livestock holdings. The size of his cattle herd has increased dramatically from an estimated 40 head in 1978 to over 100 in 1981, with two-thirds of the increase through purchases. This producer is intent upon eventually buying a pickup truck, perhaps in partnership with other nearby producers, to take advantage of the spread between prices paid for his crops at home and at markets. For example, cabbages which sold for approximately Ksh 1.80 a head at the homestead fetched as much as Ksh 5.00 a head at markets within a two-hour drive. A pickup truck is the largest expected investment for the near future. Medium-term production plans include continued cropping for sale and for home consumption, fencing of the cultivated fields, and upgrading of cattle using Sahiwal-crossed bulls. The head of WNP 1 is a progressive small-scale producer, who operates in the context of the traditional pastoral system yet, is incorporating cropping and livestock trading activities to take advantage of marketing opportunities.

WNP 2

The second small-scale unit is also located on trust land, less than 2 km from WNP 1. Two brothers, both in their mid-twenties, support five dependents. Livestock management and cropping responsibilities fall upon the younger brother, since the older one is employed in Nairobi. WNP 2 is more typical of small-scale units in the sample area than WNP 1, with a much lower frequency and volume of livestock transactions and crop sales. Over a two-year period (1980-1981), only five head of cattle were purchased expressly for resale, and cultivation of crops--maize and beans--is almost entirely for home consumption.

WNP 2's cattle herd has been expanding, to 160 head in 1981 from about 115 head in 1978. However, whereas most of WNP 1's herd growth was by purchase, the increase for WNP 2 has been mainly due to home-bred calves. Slowly, the traditional management objectives of maximizing herd size and milk offtake, and the cultivation of crops solely for home consumption, are acquiring an added commercial aspect on this production unit. A small number of steers are being purchased for fattening and resale, and beans are beginning to be sold regularly each year. WNP 2 is a small-scale operation becoming involved in marketing activities, but at a more cautious pace than WNP 1.

WNP 3

The third small-scale production unit in the sample is found about 20 km from WNP 1 and 2. In the case of WNP 3, three of four brothers have placed their herds in the care of the fourth. The three brothers are employed away from home, necessitating this arrangement. The livestock of WNP 3 are kept at three separate inkan'gitie, an arrangement notably different from WNP 1 and 2. Enkang A is located on trust land while the other two inkan'gitie, B and C, are situated on the group ranches Emorogi and Osupukiai, respectively. The three homesteads approximate the vertices of an isosceles triangle, with B and C about 2 km apart, and each about 7 km from A.

Two of the brothers are members of Emorogi group ranch, while the other two (along with four other family members) belong to Osupukiai group ranch. Both group ranches were established in 1976. Though only the one brother is presently living in the area, all four brothers intend to establish permanent homes on the group ranches in the future. At enkang A, there are eight dependents, including six children. Seventeen dependents (11 children) live at enkang B, and six dependents (four children) at enkang C. Additional family members living at the three homesteads who are not under the guardianship or care of the fourth brother have been excluded from this count. Even so, WNP 3 clearly encompasses a much more extensive family unit than does WNP 1 or 2. In the management of herds, there is frequent intermingling of stock between inkan'gitie, depending upon available grazing and the management needs of particular animals.

WNP 3 is involved to a very limited extent in market production. No animals were purchased or sold during 1981. The residing brother grows maize primarily for home consumption and beans for the market, on a slightly larger scale than the head of WNP 2. Women at each of the homesteads cultivate small plots of maize and beans using non-Maasai labor.

WNP 4

Turning to the first of the two individual ranches, WNP 4 is a 370 ha holding established in the mid-1970s on former trust land. The owner-manager is taking a methodical, step-by-step approach in the development of the ranch, recognizing that gradual, directed change is the factor lending greatest likelihood to the ranch's eventual success. This approach is apparent in his development strategy, as summarized here:

(i) Purchase of mature steers for immediate finishing and sale, and immatures, either for two-year grazing or for resale as conditioned immatures to other individual ranchers in Kajiado District. The eventual target number is 200 adult steers and 400 immatures, with grazing of immatures outside the ranch boundaries as long as there is plenty of forage and no undue disease or stock theft risks. Purchase and sale of steers is viewed as a temporary activity, intended to provide income for building up an improved breed dairy herd. With this goal in mind, the proportion of the total cattle herd composed of heifers and cows will be increased each year and the number of steers decreased.

(ii) Expansion of the breeding herd to about 300 head. Then, reduction to the best of 200 Sahiwal-crossed cows through selected culling over three to five years.

(iii) Expansion of the present flock of Dorper-crossed sheep, to 1,400 head over seven years, with mutton production the objective. Following this expansion, replacement of Dorper rams with Hampshires, the offspring of which will provide a trial testing of the ability of grade wool sheep to survive in the area, as well as animals for training workers in the production of fat lambs and high-grade mutton.

(iv) Gradual introduction of the Corriedale breed after four years, with the sale of Dorper-Hampshire crosses expected to cover purchase costs. Starting with a Corriedale flock of about 500 ewes bred with Corriedale and Hampshire rams, the breeding flock will be increased to about 1,000, a number that will then be maintained through selection and culling.

The introduction of a flock of Corriedale and Corriedale-Hampshire at a pace which will ensure that conditions suitable for such breeds exist exemplifies WNP 4's carefully planned development. Moreover, modifications of plans have been made, demonstrating management flexibility. For example, the upgrading of breeding stock is proceeding more slowly than originally intended, due to a high death loss from disease. Instead, the owner of WNP 4 is preceding the upgrading of the breeding herd by the introduction of Sahiwal-crossed steers, 20 at a time. Concerning steer feeding, the owner of WNP 4 has found that his net return is greater if immatures are grazed for only about three months and then sold for finishing to other ranchers nearer Nairobi, than if he grazes them for a year or longer.

Large investments have been made for the highly productive livestock enterprise envisaged. Long-term development loans and working capital loans from the AFC have been used for construction on the ranch, as well as to cover steer purchases and operating expenses. Medium-term loans from the Kenya Farmers Association have enabled the purchase of a tractor and farm implements. The investment in cropping machinery is for cultivating leased land outside the ranch. The owner of WNP 4 contracted with barley growers for the ploughing, harrowing, and spraying of about 88 ha for the first time in February, 1981, and earned a net return of about Ksh 400 per ha.[2] He expects his contract ploughing enterprise to expand, thus providing another source of funds for the development of WNP 4. WNP 4 also receives an income from the grazing and dipping of other producers' livestock, since it has the only properly maintained dip in the area other than that belonging to WNP 5. In 1981 this income totaled over Ksh 22,000. Table 5.27 lists the fees charged, and, as an example of earnings, Table 5.28 gives the costs for a nearby group ranch from mid-May to mid-July, 1979. A market oriented development strategy, ability to assume moderate risks, recognition of the necessity for change to be introduced gradually, willingness to modify plans as circumstances require, and a range of income earning enterprises--these stand out as reasons for WNP 4's success to date and expected continued progress.

WNP 5

This ranch borders WNP 4 and was also recently established on former trust land. However, WNP 5's

Table 5.27. Livestock management fees charged by WNP 4, 1981

Item	Charge
	--Ksh per head--
Grazing and night corraling	
Fifteen days or less	0.30 per day
Over fifteen days	5.00 per month
Dipping	
Resident livestock belonging to neighbors:	
Cattle, adult	0.40 per dipping
Calves and small stock	0.20 per dipping
or	
All livestock	20.00 per year
Transitory livestock	0.60 per dipping
Salt, medicines and other management inputs	40.00 per year

Table 5.28. Typical charges assessed a nearby group ranch by WNP 4, mid-May to mid-July, 1979

Item	Charge
	--Ksh--
Grazing and corraling of livestock	1,656.00
Dipping	328.80
Provision of salt and medicines	415.00
Labor	600.00
Total	2,999.80

development is unconstrained by financial limitations. The owner is independently wealthy, and has been able to make major capital investments. Over Ksh 150,000 was spent during the first three years in the construction of a cattle dip, a night enclosure for cattle, a calf house, a weighing house and weighbridge, and other major investments, including 6,000 poles purchased for fencing of the ranch's perimeter and grazing paddocks. Despite such large expenditures, as of 1981 the ranch had not yet produced an income. The first livestock were purchased in 1980, and none had been marketed by the end of 1981. Consequently, WNP 5 offers even less insight regarding marketing practices than does WNP 4. Still, the short history of the ranch is instructive regarding the direction development can take when funds are not a constraint. In Chapter VIII conclusions on the production, investment, and marketing potentials of WNP 4 and 5 are augmented by data from a fully operational commercial ranch located in a similar ecological setting.

Summary

The WNP sample area offers agricultural development possibilities uncommon for Kenya, namely, a semihumid environment favoring intensive land use which remains largely unexploited. Adoption of mixed farming systems is likely to gain momentum as population pressures increase, a non-Maasai labor supply remains readily available, and market returns for crops become increasingly attractive. But, for small and large-scale operations alike, it is improbable that the prevailing interest in livestock-centered enterprises will change. Producers' willingness to grow and sell crops (still anathema to many pastoral Maasai in lower-potential regions) is motivated by livestock investment possibilities.

Dairy farming is likely to become prevalent, especially on private holdings, once tenure rights are finalized and land use is intensified. Other forms of livestock production may also emerge where the management and financial capabilities exist, such as the planned sheep project on WNP 4. However, the success of all such developments will rest upon successful disease control measures. On the communally grazed group ranches in the area, producers face the same problems of lack of liability and resource controls hindering the development process on group ranches in other parts of Maasailand.

CONCLUSIONS

Principal traits of sampled Maasai producers which are considered relevant in understanding production constraints have been presented. Major characteristics of the three samples are summarized in Table 5.29, in which it is evident that they are distinguished from one another more

Table 5.29. Characteristics of producer samples

Characteristics	Samples		
	Kajiado Group Ranch (KGR)	Kajiado Individual Ranch (KIR)	Western Narok Producer (WNP)
Composition	Seventy wealth-stratified, randomly selected households of Olkarkar, Merueshi, and Mbirikani group ranches. (Budgets only for Olkarkar and Mbirikani.)	Nine private ranches, 3 in the vicinity of the KGR sample and 6 near Kiserian, south of Nairobi.	Three small-scale production units on communal land and 2 private ranches, Uasin Gishu Location, Trans Mara Division.
Eco-climatic setting	Zone V	Zones IV and V	Zone III
Production orientation	Noncommercial	Mixed	Mixed
Capital investment	Negligible	Mixed	Mixed
Cultivation of crops	No	No	Yes

by tenurial status and eco-climatic setting than by actual livestock management practices. Relative progressiveness of production practices does not appear to vary strictly by sample. The disparity in production orientations between units within each of the samples attests to the futility of attempting to generalize only by sample. Constraints to the transition process, in particular ineffective resource controls, require that attention be given to the individual units in the KIR and WNP samples, and to differences between the KGR sample households by location and wealth stratum. The micro-level examination that has been qualitatively initiated in this chapter is quantitatively pursued in Chapter VI by means of budgetary analysis.

NOTES

[1] The total "livestock equivalent" (LE) value for a household is based upon the size of its herd/flock holdings, adjusted by a measure of the animals' metabolic weights: ($liveweight^{0.75}$). Employing average liveweights reported in studies at Elangata Wuas group ranch by Semenye [1980], namely, 122 kg for cattle with milk teeth, 255 kg for those with two to four adult incisors, and 331 kg for cattle with six to eight adult incisors, metabolic weights of 36.70, 63.81, and 77.60 were computed for the three age groups. Assigning 1 LE to an adult bovine, an animal with two to four adult incisors is therefore represented by 0.82 LE, and an animal with milk teeth, by 0.47 LE. Overall herd structures on the three ranches were estimated as 48 percent of the cattle with milk teeth, 17 percent with two to four adult incisors, and 35 percent with six to eight incisors. Multiplying each of these percentages by its LE value yields the following coefficients: 0.22 for cattle with milk teeth, 0.14 for cattle with two to four adult incisors, and 0.35 for cattle with six to eight adult incisors, or an aggregate coefficient of 0.71. Multiplying the number of cattle in a household's herd by 0.71 yields its total cattle LEs. An aggregate coefficient of 0.17 was derived in the same manner for small stock, using average liveweights reported on Elangata Wuas group ranch by Wilson, Peacock, and Sayers [1981]. A household's total small stock LEs was determined by multiplying the number of sheep and goats by 0.17.

The "active adult male equivalent" (AAME) represents the average daily food energy requirements of an active African adult male based upon FAO recommendations, that is, 2,530 kcal [FAO 1974]. Other members of the population are counted as follows: adult female, 0.86 AAME; child, 0-5 years, 0.52 AAME; child, 6-10 years, 0.85 AAME; and child, 11-15 years, 0.96 AAME. Total AAMEs for a household was determined by its size and composition.

[2]Contract ploughing and cropping on leased land has been practiced extensively for the last 15 years in some parts of Narok District, as described in Chapter IV. Barley growing is usually financed by Kenya Breweries Limited, and AFC often handles wheat, delivering inputs to the site, supervising planting, spraying, and harvesting, and arranging for temporary storage and crop transportation. The owner of WNP 4 planned to contract for both wheat and barley. While barley is higher yielding than wheat and susceptible to fewer diseases, its grading system is apparently more difficult to predict and prices are lower.

Maize, pyrethrum, potatoes, beans, tobacco, and sugarcane also hold potential for contract cultivation in Trans Mara Division. No major contract ploughing had taken place in the division as of 1981, even though the land leasing rate was Ksh 60 to Ksh 80 per acre, compared to about Ksh 160 per acre in areas of Narok District regularly cultivated. The owner of WNP 4 intends to expand his present operation from two tractors to five (with implements), to take advantage of opportunities in the WNP sample area.

6
Production Costs and Returns of Sampled Units

The descriptions of producer samples given in the preceding chapter set the stage for quantitatively analyzing the various livestock production units in order to better assess constraints to the pastoral transition process. Budgeting is the analytical method employed. Summary information from the production budgets is presented, and the results of White and Meadows' [1981] study of Kajiado individual and group ranches are used to verify cost-and-return values and offtake rates determined for the KGR and KIR samples. Production-marketing relationships are then investigated in the second section of the chapter.

PRODUCTION UNIT BUDGETS

Budgeting as a Tool of Analysis

The use of budgets in operating a commercial farm or ranch is well recognized as an activity integral to profitable decision making. Budgeting is most effectively utilized as an ongoing process which takes into account changes over time in resource availability, market and nonmarket values of inputs and products, and producers' objectives [Rae 1977]. The usefulness of budgets in studying agricultural systems, although not widely appreciated, is well demonstrated by Simpson and Farris [1982]. Contrasting budgets representative of cattle production systems in southern Texas, Paraguay, and Tanzania, they were able to dramatically highlight differences among the three systems. Budgeting as an analytical technique is employed here in this manner, in order to define and compare costs and returns of Maasai production units, and thereby better identify constraints to expanded livestock production.[1]

Budgets have been developed for each production unit of the Kajiado Individual Ranch (KIR) and Western Narok Producer (WNP) samples. Six budgets (representing the three wealth strata on the two group ranches, Olkarkar and

Mbirikani) have been constructed for the Kajiado Group Ranch (KGR) sample. As described in Chapter V, the statistical representativeness of the group ranch budgets permits conclusions to be stated with greater confidence than is the case with the KIR and WNP samples. The analytical strength of the latter budgets lies in the precision with which individual units are depicted.

Principal Characteristics of the Budgets

The KGR sample. The statistics derived from the KGR budgets are summarized in Table 6.1. For all six stratum-ranch combinations, incomes above production costs were found to be positive, but incomes above total costs, that is, production plus opportunity costs, are negative, as is true for all three samples. Distinctions across strata and between ranches are readily apparent. The wealthier households (Stratum III) have higher levels of sales, but sales are greater as a proportion of gross incomes--and per AU--for the poorer households. Gross income per ha and sales per ha are estimated to be higher for Olkarkar than for Mbirikani. Assumed production costs of Ksh 20 per AU imply operating costs of Ksh 8 per ha for Olkarkar and Ksh 5 per ha for Mbirikani. These findings corroborate production-marketing relationships indicated in Chapter V, namely (i) livestock sales are positively correlated with wealth (livestock holdings) yet sales, as a proportion of total income, are negatively correlated with wealth, and (ii) Olkarkar group ranch, more intensively grazed and nearer marketing centers than Mbirikani, exhibits higher levels of production per ha.

The KIR sample. In a similar manner, the KIR budgets underscore the disparity among individual ranches sampled (Table 6.2). For example, capital and equipment investments range from Ksh 26 per AU for both KIR 2 and KIR 3, to Ksh 932 per AU for KIR 4. The prominence of slaughter stock production is evident, with only KIR 6 having a higher gross income from milk than from meat production. Also noteworthy is the major share of total income which is earned from the sale of AFC-financed steers by the four ranches which sold purchased steers during the survey year. As with the KGR budgets, incomes above production costs are positive (except for KIR 4, KIR 6, and KIR 9, for which large livestock sales following the survey period were not included), and total net incomes are negative.

The apparently poor showing by KIR 9 illustrates a serious shortcoming in relying solely upon single-period budgeting to access a production unit's profitability. KIR 9's income for the survey year was negative, since no cattle other than culled cows were sold. Yet, as described in Chapter V, KIR 9 is clearly the most progressive ranch in the sample. Table 6.3 compares the ranch's actual costs and returns with those which would

Table 6.1. Summary statistics on costs and returns per A.U., by stratum, Kajiado Group Ranch Sample

Item	Unit	Olkarkar Group Ranch				Mbirikani Group Ranch			
		Stratum I	Stratum II	Stratum III	Ranch	Stratum I	Stratum II	Stratum III	Ranch
		Ksh per A.U.							
Capital and equipment investment[a]		40	15	6	20	25	11	6	14
Livestock investment		1,137	1,093	1,284	1,168	1,219	1,208	1,240	1,233
Production costs		32	28	30	30	32	31	31	31
Opportunity costs		384	225	191	262	302	215	186	202
Gross income, including home consumption		201	245	206	219	273	223	188	201
Income above production costs		169	217	176	189	241	192	157	170
Income above production and opportunity costs		-183	-8	-15	-73	-61	-23	-29	-32
Proportion of gross income									
from meat	%	62	61	48	--	60	52	43	--
milk	%	38	39	52	--	40	48	57	--
Sales, value	Ksh	3,365	9,207	15,405	--	7,098	9,529	14,102	--
as a proportion of gross income	%	56	47	36	--	54	39	36	--
per A.U.	Ksh	112	115	74	101	148	87	67	78

[a]Not including Olkarkar's AFC-financed dip and piping system.

Table 6.2. Summary statistics on costs and returns, per ha and per A.U., Kajiado Individual Ranch sample

Item	Unit	KIR 1	KIR 2	KIR 3	KIR 4	KIR 5	KIR 6	KIR 7	KIR 8	KIR 9
On-ranch grazing area	ha	800	560	680	158	1,440	237	240	480	720
Total grazing area utilized	ha	1,013	840	1,020	658	1,700	474	480	480	1,200
Grazing pressure, if only the										
ranch were used	ha per A.U.	3.0	1.8	5.5	0.5	1.8	1.1	0.9	...	1.3
actual	ha per A.U.	3.8	2.6	8.2	2.1	2.1	2.1	1.8	...	2.2
Capital and equipment investment										
per ha, ranch only	Ksh	68	15	5	1,882	75	772	427	368	628
actual	Ksh	54	10	3	452	64	386	214	368	377
per A.U.	Ksh	204	26	26	932	133	824	381	...	815
Livestock investment										
per ha, ranch only	Ksh	482	906	293	3,108	1,295	1,590	1,668	688	1,931
actual	Ksh	380	604	195	746	1,097	795	834	688	1,158
per A.U.	Ksh	1,443	1,601	1,608	1,539	2,291	1,697	1,488	...	2,505
Production costs										
per ha, ranch only	Ksh	174	25	172	899	200	243	593	178	480
actual	Ksh	138	17	115	216	169	121	296	178	288
per A.U.	Ksh	522	45	943	445	353	259	529	...	623
Opportunity costs										
per ha, ranch only	Ksh	1,068	1,103	1,042	8,439	6,203	6,394	5,574	6,176	6,403
actual	Ksh	844	736	695	2,026	5,254	3,197	2,787	6,176	3,841
per A.U.	Ksh	3,201	1,949	5,715	4,180	10,973	6,826	4,973	...	8,306
Gross income, including home consumption										
per ha, ranch only	Ksh	220	191	284	577	280	122	873	...	112
actual	Ksh	173	127	189	139	237	61	437	...	67
per A.U.	Ksh	658	337	1,558	286	495	130	779	...	145
Proportion of gross income from										
livestock (meat)	%	85	77	96	78	89	7	79	...	40
milk	%	15	23	4	22	11	93	21	...	60

Table 6.2. continued

Item	Unit	KIR 1	KIR 2	KIR 3	KIR 4	KIR 5	KIR 6	KIR 7	KIR 8	KIR 9
Income above production costs										
per ha, ranch only	Ksh	46	166	112	-322	80	-121	280	...	-368
actual	Ksh	35	110	74	-77	68	-60	141	...	-221
per A.U.	Ksh	136	292	615	-159	142	-129	250	...	-478
Income above production and opportunity costs										
per ha, ranch only	Ksh	-1,022	-937	-930	-8;761	-6,123	-6,515	-5,294	...	-6,771
actual	Ksh	-809	-626	-621	-2,103	-5,186	-3,257	-2,646	...	-4,062
per A.U.	Ksh	-3,065	-1,657	-5,100	-4,339	-10,831	-6,955	-4,723	...	-8,784
Sales										
value	Ksh	140,000	57,800	176,000	68,000	343,400	--	165,500	...	28,750
as a proportion of gross income	%	80	54	91	75	85	--	79	...	36
AFC-financed steers										
value	Ksh	140,000	--	160,000	68,000	--	--	112,000	--	--
as a proportion of total sales	%	100	--	91	100	--	--	68	--	--
Per ha, ranch only	Ksh	175	103	259	430	238	--	690	...	40
actual	Ksh	138	69	173	103	202	--	345	...	24
Per A.U.	Ksh	524	182	1,419	213	422	--	615	...	52

Table 6.3. Comparison of KIR 9's actual returns for 1981 with returns if steers sold in February, 1982, had been sold in December, 1981

Item	Costs and Returns	
	Actual	If steers were sold in December
	----------Ksh----------	
Production costs	345,876	345,876
Opportunity costs	4,610,069	4,610,069
Gross income		
Cattle, sales	---	673,200[a]
home consumption	48,600	48,600
Small stock, sales	28,750	28,750
home consumption	3,000	3,000
Total	80,350	753,550
Income above production costs	-265,526	407,674
Income above production and opportunity costs	-4,875,595	-4,202,395

[a] 340 steers x 330 kg/steer x Ksh 6 per kg = Ksh 673,200.

have resulted if the 340 steers had been sold in December, 1981, instead of February, 1982. In the former case, income above production costs would have been positive. Ideally, such misrepresentations are overcome by the use of multi-period budgeting. When only single-period budgeting is possible, qualitative background information as provided in Chapter V is essential to accurately perceive a unit's operation.

The WNP sample. The distinction between the small production units, WNP 1, 2, and 3, and the larger private holdings, WNP 4 and 5, is dramatically apparent in their levels of investment and production costs (Table 6.4). Only the small producers have positive incomes above production costs, due to the early stage of development of WNP 4 and 5. In contrast to the Kajiado samples, only one production unit, WNP 2, recorded any livestock sales during the survey period (although the head of WNP 1 engaged in extensive cattle trading, described in Chapter VIII). As with KIR 9, a single-year budget misrepresents WNP 4's active history of cattle transactions discussed in Chapter V and summarized in Table 6.5. Crop budgets show that the cultivation of subsistence crops, principally maize, is a major activity of the small-scale units. Marketing of crops by the head of WNP 1, in particular, demonstrates their income earning potential.

Verifying Measures for the KGR and KIR Samples

It is appropriate to consider whether the budgets realistically represent the units/strata before proceeding to a discussion of production-marketing relationships. One excellent comparative basis for the KGR and KIR budgets is provided by White and Meadows' [1981] analysis of Kajiado group and individual ranches. White and Meadows' [1981] study, denoted W&M, is based on data gathered from 60 households on five group ranches (12 on each ranch), and 23 individual ranches located in four different areas of the district, from August, 1980, to July, 1981. Range conditions were unfavorable during the first eight months of data collection, while for the last four months they were very good. Comparisons between W&M's findings and incomes, expenditures, and offtake rates for the KGR and KIR samples are summarized in Table 6.6.

Beginning with expenditures, there was a widespread dependence upon purchased maize during the dry season by W&M group ranch households. Sugar, beer, and meat were items also frequently purchased. Annual per capita expenditures on food and drink items by the 60 group ranch households ranged from Ksh 278 to Ksh 473, comparing closely with consumption expenditures reported for Olkarkar and Mbirikani households (Table 5.12).

Livestock production expenditures overshadowed consumption expenditures on both W&M group and individual ranches. Operating costs for group ranch households were

Table 6.4. Summary statistics on livestock costs and returns, per ha and per A.U., Western Narok Producer sample

Item	Unit	WNP1	WNP2	WNP3	WNP4	WNP5
Capital and equipment investment						
Per ha	Ksh	NA[a]	NA	NA	1,025	793
Per A.U.	Ksh	12	10	7	726	3,883
Livestock investment						
Per ha	Ksh	NA	NA	NA	1,908	247
Per A.U.	Ksh	625	794	784	1,352	1,212
Production costs						
Per ha	Ksh	NA	NA	NA	725	320
Per A.U.	Ksh	77	67	23	514	1,569
Opportunity costs						
Per ha	Ksh	NA	NA	NA	1,380	1,075
Per A.U.	Ksh	172	167	171	977	5,266
Gross income, including home consumption						
Per ha	Ksh	NA	NA	NA	163	--
Per A.U.	Ksh	106	184	70	116	--
Proportion of gross income from						
Livestock (meat)	%	8	25	4	3	--
Milk	%	92	75	96	97	--
Income above production costs						
Per ha	Ksh	NA	NA	NA	-562	-320
Per A.U.	Ksh	29	117	47	-398	-1,569
Income above production and opportunity costs						
Per ha	Ksh	NA	NA	NA	-1,942	-1,395
Per A.U.	Ksh	-143	-50	-124	-1,375	-6,835
Sales, value	Ksh	--	300	--	--	--
as a proportion of gross income	%	--	1	--	--	--
Per ha	Ksh	NA	NA	NA	--	--
Per A.U.	Ksh	--	2	--	--	--

[a]Not applicable.

Table 6.5. Cattle transactions, WNP4, mid-1978 to mid-1981

	Purchase			Sale		
		Price			Price	
Date	Quantity and Sex[a]	Mean	Total	Quantity and Sex	Mean	Total
		-Ksh per head-	-Ksh-		-Ksh per head-	-Ksh-
June, 1978	74 steers	629	46,255			
October, 1978	35 steers	1,022	35,765			
February, 1979	80 heifers	625	50,000			
July, 1979				90 steers	1,500	135,000
July, 1979	150 steers	900	135,000			
August, 1979				60 steers	1,300	78,000
September, 1979	50 heifers and 50 steers	570	57,000			
February, 1980				42 steers	1,200	50,400
March, 1980				80 steers	1,000	80,000
April, 1980	120 steers	700	84,000			
June, 1981	150 steers	600	90,000			

[a]Most frequently purchased as a herd at a price per head. Ages of purchased cattle generally ranged from three to five years.

Table 6.6. Comparison of values for KGR and KIR budgets with values from White and Meadows [1981] sample of Kajiado group and individual ranches

		Sample			
				White and Meadows	
Item	Unit	KGR	KIR	Group Ranches	Individual Ranches
		mean annual values			
Survey period	--	July, 1981-June, 1982	Jan, 1981-Dec, 1981	--- August, 1980-July, 1981 ---	
Livestock purchases	Ksh per household	650-1,700	none-181,800	1,304-8,554	2,337-84,626
Direct production expenditures	Ksh per household	600-4,200[a]	10,545-99,740[c]	1,236-7,652	6,113-52,602[c]
Consumption expenditures on food and drink items	Ksh per capita	290-361[b]	--	278-473	--
Total cash income	Ksh per capita	775	--	1,568	--
Proportion of total income from sales	percent	47	70	57	80
Cattle commercial offtake rates (sales)	percent	10	25	11	21

[a]Estimated for budgets.

[b]Adjusted from Ksh per AAME by multiplying by a factor of 0.8.

[c]Excluding capital and equipment maintenance and improvements.

not great, and livestock purchases and production expenditures for the W&M individual ranches reflect levels of market involvement similar to that of the KIR sample ranches. Mean production expenditures per capita for W&M group ranch households were larger than for the KGR sample, while costs for W&M individual ranches were less than for the KIR sample ranches. Still, differences between corresponding samples are within a reasonable range.

Mean cash incomes per capita for W&M group ranch households is about twice that of the KGR sample. On four of the five W&M group ranches, cash incomes exceeded subsistence incomes, a condition which held only for Stratum I households of the KGR sample. Notably, however, milk is valued at only Ksh 1.40 per liter in White and Meadows' study, compared to Ksh 2.50 in the KGR budgets. Cattle sales (including interhousehold sales) ranged from 9.2 to 12.3 percent for W&M group ranch households, and nonsale offtake, from 7.7 to 10.6 percent. The W&M individual ranches had sales rates, ranging from 18.8 to 22.6 percent. Comparable offtake rates hold for the KGR and KIR samples (Table 6.6). Without the steer-feeding enterprises described in Chapter V, herd structures for individual ranches would differ little from those of Maasai households on group ranches. This fact is supported by both the KIR and W&M samples, as shown in Table 6.7. The proportions of herds which are cows drops below 50 percent and numbers of steers practically double when steers purchased for grazing are included in inventories. Nonsale offtake rates for cattle on the W&M individual ranches were low, similar to the percentages for the group ranch households, and consisted mainly of culled cows and calves. In sum, rates and magnitudes reported for the W&M group and individual ranches lend verifying support to the values derived from the KGR and KIR budgets.

An additional check on the representativeness of the KGR budgets comes from information gathered from 19 producers interviewed by the author in 1980 at Emali, the principal livestock marketing center for the area. The relative frequencies of their commercial transactions during the previous year are shown in Table 6.8. Numbers of animals involved in the transactions are not available, but most livestock are sold singly by Maasai producers, or in groups of two or three. Generally, the mean frequency for each type of transaction--purchase/sale, of cattle/small stock, at home/market--was reported to be three to five times during the year, with sales more common than purchases. Cattle sales at the market were by far the most common commercial transaction, small stock sales at the market the least common, and purchases and sales at home of all species of intermediate occurrence. Notwithstanding the high level of small stock sales by

Table 6.7. Composite cattle herd structures for KIR sample ranches, 1982, and 23 individual ranches sampled by White and Meadows [1981], Kajiado District, 1980-1981

Type	KIR Sample				White and Meadows Sample[a]			
	Including Purchased Steers		Excluding Purchased Steers		Including Purchased Steers		Excluding Purchased Steers	
	-head-	-percent-	-head-	-percent-	-head-	-percent-	-head-	-percent-
Cows over one year	1,374	40	1,374	54	3,183	44	3,183	60
Steers and bulls over one year	1,619	47	730	28	3,071	43	1,158	22
Calves	450	13	450	18	928	13	928	18
Total	3,443	100	2,554	100	7,182	100	5,269	100

[a]White and Meadows [1981], Table 8.

Table 6.8. Frequency of commercial transactions for 19 producers in the KGR sample area

Number of Instances During the Previous Year[a]	Type and Place of Transaction							
	Purchases				Sales			
	At Home		At Market		At Home		At Market	
	Cattle	Small Stock	Cattle	Small Stock	Cattle	Small Stock	Cattle	Small Stock
	---number of producers--- (n = 19)							
1-2	2	4	1	4	1	3	7	1
3-5	4	6	6	6	4	2	8	2
6-10	-	-	-	2	4	4	2	2
more than 10	-	-	-	-	-	5	1	2

[a]Does not distinguish between differences in the numbers of animals involved in the transactions, but almost all purchases and sales were of single animals, and secondarily, of groups of two or three.

Olkarkar households, these patterns of commercial offtake agree with those indicated in the KGR budgets.

Although there are no secondary sources by which to likewise check the WNP budgets, neither is there cause to suspect that the values are unrealistic. The same procedural steps were followed in the construction of the WNP budgets as were taken for the KGR and KIR budgets. Having presented the budgets, briefly highlighted their principal characteristics, and corroborated values using secondary sources, production-marketing relationships and development options are now considered.

PRODUCTION-MARKETING LEVELS AND RELATIONSHIPS

The principal purpose of the budgets is to determine livestock production and marketing magnitudes and relationships for the sampled systems. Relative magnitudes have been discussed in the preceding section, and the means and standard deviations of various production and marketing measures for the three samples are shown in Table 6.9. Gross incomes are greater for the individual ranch sample than for the group ranch sample. Net incomes on Olkarkar are more comparable to those of the individual ranches, due to the latter's higher production costs. The high mean production costs and negative mean net incomes for the WNP sample derive from the sizable expenditures and absence of sales by the two private holdings, WNP 4 and 5. Finally, the effects on sales levels for the KIR sample units of including or excluding (i) use of off-ranch grazing and (ii) AFC-financed steers are evident. For example, the use of land outside ranch boundaries reduces sales per ha for the KIR sample by 45 percent, from Ksh 242 to Ksh 132. If sales of AFC-financed steers are not included in the calculations, sales as a proportion of gross income for the sample decline from 63 percent to 26 percent. The high standard deviations of the measures reflect the wide range in values recorded for the KIR sample units.

Hypotheses stated here are used to guide the examination of production-marketing relationships. Gross incomes and net incomes (that is, incomes over net production costs), per ha and per AU, serve as proxy measures of average productivity in testing the hypotheses. For the KIR sample, paired average productivities per ha are presented, as in Table 6.2, since almost all of the individual ranches utilized off-ranch grazing. With no data collected on KIR 8's livestock holdings, the KIR sample is reduced to eight ranches. Data used to test the hypotheses are found in Table 6.10.

Hypothesis 1. Average productivities per livestock unit (AU) for private and group owned ranches in similar eco-climatic settings in Maasailand do not differ significantly.

Table 6.9. Mean value and standard deviation of production and marketing measures for the KGR, KIR, and WNP samples

Item	Unit	KGR Sample		KIR Sample		WNP Sample	
		Mean	Std Dev	Mean	Std Dev	Mean	Std Dev
Gross income, per A.U.	Ksh	223	32	549	469	95	67
per ha, ranch only	Ksh	...[a]	...	332	263	...	...
per ha, total	Ksh	...	...	179	120	...	...
Net income, per A.U.	Ksh	203	32	234	273	-181	415
per ha, ranch only	Ksh	...[b]	...	107	172	...	...
per ha, total	Ksh	...	...	73	93	...	...
Primary production costs, per A.U.	Ksh	...[c]	...	396	234	278	364
per ha, ranch only	Ksh	...[d]	...	291	247	...	...
per ha, total	Ksh	...	...	143	77	...	...
Sales, as a proportion of gross income	%	45	9	63	31	...	...
per A.U.	Ksh	101	30	428	456	...	...
as a proportion of gross income, not including AFC-financed steers	%	...	...	26	31	...	...
per A.U., not including AFC-financed steers	Ksh	...	...	153	156	...	...
per ha, ranch only	Ksh	...	...	242	227	...	...
per ha, total	Ksh	...	...	132	111	...	...
per ha, ranch only, not including AFC-financed steers	Ksh	...	...	79	100	...	...
per ha, total, not including AFC-financed steers	Ksh	...	...	53	72	...	...

[a]Gross income for Olkarkar, Ksh 91 per ha; for Mbirikani, Ksh 53 per ha.

[b]Net income for Olkarkar, Ksh 83 per ha; for Mbirikani, Ksh 48 per ha.

[c]Assumed primary production costs in budgets: Ksh 20 per A.U.

[d]Given assumed production costs, for Olkarkar, Ksh 8 per ha; for Mbirikani, Ksh 5 per ha.

Table 6.10. Data on KGR, KIR, and WNP samples

Production Unit/Group Ranch Stratum	Gross Income per A.U.	Gross Income per ha, Ranch Only	Gross Income per ha, Total	Net Income per A.U.	Net Income per ha, Ranch Only	Net Income per ha, Total	Primary Production Costs per A.U.	Primary Production Costs per ha, Ranch Only
	Ksh							
Kajiado Group Ranch Sample								
Olkarkar, stratum I	201	--	--	181	--	--	--	--
stratum II	245	--	--	225	--	--	--	--
stratum III	206	--	--	186	--	--	--	--
Mbirikani, stratum I	273	--	--	253	--	--	--	--
stratum II	223	--	--	203	--	--	--	--
stratum III	188	--	--	168	--	--	--	--
Kajiado Individual Ranch Sample[a]								
KIR 1	658	220	173	278	93	73	380	127
KIR 2	337	191	127	351	199	133	125	71
KIR 3	1,558	284	189	698	127	85	860	157
KIR 4	286	577	139	-52	-106	-25	388	784
KIR 5	495	280	237	293	166	140	349	197
KIR 6	130	122	61	131	123	62	121	113
KIR 7	779	873	437	358	401	201	433	486
KIR 9	145	112	67	-188	-145	-87	513	395
Western Narok Producer Sample								
WNP1	106	--	--	36	--	--	70	--
WNP2	184	--	--	139	--	--	53	--
WNP3	70	--	--	54	--	--	15	--
WNP4	116	--	--	-262	--	--	380	--
WNP5	0	--	--	-874	--	--	874	--

Table 6.10. continued

	Primary Production Costs per ha, Total	Sales as a Proportion of Gross Income	Sales per A.U.	Sales as a Proportion of Gross Income, Excluding Purchased Steers	Sales per A.U., Excluding Purchased Steers	Sales per ha, Ranch Only	Sales per ha, Total	Sales per ha, Ranch Only, Excluding Purchased Steers	Sales per ha, Total, Excluding Purchased Steers
	--Ksh--	-percent-	--Ksh--	-percent-	----------Ksh----------				
Olkarkar, I	--	56	112	56	112	--	--	--	--
II	--	47	115	47	115	--	--	--	--
III	--	36	74	36	74	--	--	--	--
Mbirikani, I	--	54	148	54	148	--	--	--	--
II	--	39	87	39	87	--	--	--	--
III	--	36	67	36	67	--	--	--	--
KIR 1	100	80	524	0	0	175	138	0	0
2	47	54	182	54	182	103	69	103	69
3	104	91	1,419	8	296	259	173	24	16
4	188	75	213	0	0	430	103	0	0
5	167	85	422	85	422	238	202	238	202
6	57	0	0	0	0	0	0	0	0
7	243	79	615	25	218	690	345	223	111
9	237	36	52	36	102	40	24	40	24
WNP1	--	--	--	--	--	--	--	--	--
2	--	--	--	--	--	--	--	--	--
3	--	--	--	--	--	--	--	--	--
4	--	--	--	--	--	--	--	--	--
5	--	--	--	--	--	--	--	--	--

[a]KIR 8 excluded due to insufficient data.

Note: Dash implies not applicable or insufficient data.

The KIR and KGR samples are used to test this hypothesis. The degree of similarity between their physical settings may be questioned, since as described in Chapter V, the production units comprising the KIR sample--other than KIR 1, 2, and 3--lie in a generally less arid area than the KGR sample. Moreover, Olkarkar and Mbirikani sub-samples could be considered eco-climatically distinct, despite both lying within Zone V. Nonetheless, the ecological conditions that prevail are considered sufficiently homogeneous for a valid test.

Analysis of variance is employed to test whether the group and individual ranches differ significantly in their average productivities per AU. The F-ratio test statistic has 1 and 12 degrees of freedom, and the critical value is 4.75, assuming a 0.05 level of significance:

$$F = \frac{\text{Mean Sum of Squares Between Samples}}{\text{Mean Sum of of Squares Within Samples}}$$

$$= \frac{\text{Sum of Squares Between}/m - 1}{\text{Sum of Squares Within}/n - m} \qquad m = 2,\ n = 14$$

$$\text{Level of significance} = 0.05 = P(F_{1,12} > 4.75)$$

Hypothesis 1a. H_0: Gross income per AU for individual ranches = Gross income per AU for group ranch producers. H_A: Gross income per AU for individual ranches ≠ Gross income per AU for group ranch producers.

F = 2.83, therefore H_0 is not rejected at the 0.05 level of significance.

Hypothesis 1b. H_0: Net income per AU for individual ranches = Net income per AU for group ranch producers. H_A: Net income per AU for individual ranches ≠ Net income per AU for group ranch producers.

F = 0.08, therefore H_0 is not rejected at the 0.05 level of significance.

Thus, there is no significant difference between the gross or net incomes per AU (proxy for average livestock productivity) for individual and group ranch producers.

<u>Hypothesis 2</u>. Private control of resources is not a sufficient condition for production to become primarily market oriented, that is, levels of marketing activity--the value of all livestock and livestock products sold in one year relative to total holdings--do not differ significantly between private and group owned ranches.

Analysis of variance is again employed to test this hypothesis by comparing the levels of marketing activity of the KGR and KIR samples. Two measures of marketing activity are used: the value of sales as a proportion of gross income, and the value of sales per AU. The critical F-value, 4.75, is the same as that used in testing Hypothesis 1.

Hypothesis 2a. H_0: Sales as a proportion of gross income for individual producers = Sales as a proportion of gross income for group ranch producers. H_A: Sales as a proportion of gross income for individual producers ≠ Sales as a proportion of gross income for group ranch producers.

F = 1.83, therefore H_0 is not rejected at the 0.05 level of significance.

Hypothesis 2b. H_0: Value of sales per AU for individual ranches = Value of sales per AU for group ranch producers. H_A: Value of sales per AU for individual ranches ≠ Value of sales per AU for group ranch producers.

F = 3.02, therefore H_0 is not rejected at the 0.05 level of significance.

Hypothesis 2c. H_0: Sales as a proportion of gross income for individual producers, excluding AFC-financed steers = Sales as a proportion of gross income for group ranch producers. H_A: Sales as a proportion of gross income for individual producers, excluding AFC-financed steers ≠ Sales as a proportion of gross income for group ranch producers.

F = 2.01, therefore H_0 is not rejected at the 0.05 level of singnificance.

Hypothesis 2d. H_0: The value of sales per AU for individual ranches, excluding AFC-financed steers = The value of sales per AU for group ranch producers. H_A: The value of sales per AU for individual ranches, excluding AFC-financed steers ≠ The value of sales per AU for group ranch producers.

F = 0.63, therefore H_0 is not rejected at the 0.05 level of significance.

In each case, Hypothesis 2 is not rejected. There is no significant difference between the levels of marketing activity on private and group ranches, implying that private control of resources is not a sufficient condition for increased commercial production.

<u>Hypothesis 3</u>. Average productivities per livestock and per land unit are positively correlated with levels of marketing activity.

A number of correlations were carried out for the KGR and KIR samples in examining this hypothesis, as shown in Table 6.11. The correlation between gross income, as well as net income, per AU and sales per AU for the KGR sample is 0.85. Correlation coefficients for the corresponding KIR sample variables are 0.99 and 0.83, respectively. In terms of productivity per unit of land for the individual ranches, strong correlations exist between gross income per ha and sales per ha (0.98, ranch land only; 0.98, all land grazed). However, net income--sales correlations are weaker (0.46, ranch land only; 0.70, total land), indicating that production costs are disporportional to incomes from sales. Once again, the distinctions between correlations including and excluding the sale of AFC-financed steers are apparent.

Table 6.11. Correlations between measures of average productivity and level of marketing activity, KGR and KIR samples

Sample	Productivity Variable	Marketing Variable	Correlation
KGR	Gross income per A.U.	Sales, as a proportion of gross income	.51
		per A.U.	.85
	Net income per A.U.	Sales, as a proportion of gross income	.51
		per A.U.	.85
KIR	Gross income per A.U.	Sales, as a proportion of gross income	.67
		per A.U.	.99
		Sales, as a proportion of gross income, not including AFC-financed steers	-.16
		per A.U., not including AFC-financed steers	.48
	Gross income per ha, ranch only	Sales, as a proportion of gross income	.48
		as a proportion of gross income, not including AFC-financed steers	-.12
		Sales, per ha, ranch only	.98
		per ha, ranch only, not including AFC-financed steers	.46
	Gross income per ha, total	Sales, as a proportion of gross income	.62
		as a proportion of gross income, not including AFC-financed steers	.18
		Sales, per ha, total	.98
		per ha, total, not including AFC-financed steers	.57
	Net income per A.U.	Sales, as a proportion of gross income	.51
		per A.U.	.83
		Sales, as a proportion of gross income, not including AFC-financed steers	.05
		per A.U., not including AFC-financed steers	.56

Table 6.11. continued

Sample	Productivity Variable	Marketing Variable	Correlation
KIR	Net income per ha, ranch only	Sales, as a proportion of gross income	.23
		as a proportion of gross income, not including AFC-financed steers	.22
		Sales, per ha, ranch only	.46
		per ha, ranch only, not including AFC-financed steers	.66
	Net income per ha, total	Sales, as a proportion of gross income	.36
		as a proportion of gross income, not including AFC-financed steers	.33
		Sales, per ha, total	.70
		per ha, total, not including AFC-financed steers	.61

Hypothesis 4. Average productivities and levels of production expenditure are positively correlated.

Correlations of gross and net incomes for the KIR and WNP samples with production costs are shown in Table 6.12. Correlations involving gross incomes are positive for the KIR sample, whereas those between net incomes and production costs--on a per ha basis--are negative, reflecting the low net incomes of ranches with highest production costs. The negative correlations obtained for the WNP sample are explained by the high production costs of the two private holdings and their negative net incomes. Given the test results for Hypothesis 1 and the negative correlations described here, Hypothesis 4 cannot be accepted. There is no basis for assuming average productivity levels are positively correlated with levels of production expenditure.

In summary, the budgeted samples suggest that (i) average livestock productivities of group and individual ranches do not differ significantly, (ii) levels of marketing activity relative to holdings for the two production systems also do not differ significantly, (iii) average productivities under both systems are positively correlated with levels of marketing activity, and (iv) units with high production costs do not necessarily have high incomes.

CONCLUSIONS

The KGR budgets depict notable differences among pastoral producers, by location and household wealth. The budgets for Olkarkar and Mbirikani group ranch households and distinctions between wealth strata indicate that each household's transition to increased commercial production will depend on its particular circumstances, particularly relative population densities in the area, nearness to markets, and size of livestock holdings. Proposals for group ranch development projects need to recognize these differences between member households for as current production practices vary so will responses to incentives to expand production.

Low levels of development for many of the KIR sample ranches are apparent from their budgets, to the point that production inputs for several of them differ little from those found on group ranches. Indeed, when AFC-financed steer-feeding enterprises are excluded, ranches in the sample which are clearly guided by market oriented production strategies are the exception. Finally, prevailing traditional production practices of small-scale producers in the WNP sample area and the significance of cultivation in this more humid region of Maasailand are revealed by the WNP budgets.

The budgets and subsequent tests of production-investment-marketing relationships support the conclusion tentatively drawn in previous chapters, namely,

Table 6.12. Correlations between measures of average productivity and primary production costs, KIR and WNP samples

Sample	Productivity Variable	Cost of Production Variable	Correlation
KIR	Gross income per A.U.	Primary production costs per A.U.	.79
	Gross income per ha, ranch only	Primary production costs per ha, ranch only	.67
	Gross income per ha, total	Primary production costs per ha, total	.46
	Net income per A.U.	Primary production costs per A.U.	.37
	Net income per ha, ranch only	Primary production costs per ha, ranch only	-.35
	Net income per ha, total	Primary production costs per ha, total	-.21
WNP	Gross income per A.U.	Primary production costs per A.U.	-.72
	Net income per A.U.	Primary production costs per A.U.	-.99

ineffective control over resource use is an overriding limiting factor in the continuum of constraints preventing expanded livestock production in Maasailand. The similarity in levels of productivity and marketing activities of sampled group and individual ranches indicates that land privatization does not automatically result in a commercialized system. Private ownership provides the opportunity for effective resource control, but only on those ranches where controls have been enforced and investments initiated can there be a basis for increased production.

Maasai producers and their practices have been the focus so far in the examination of constraints. However, inefficient marketing mechanisms may be constraining expanded production as well. The local-level trade in livestock is investigated in the following chapter.

NOTE

[1] Budget inventories, flows, production coefficients, and costs and returns are presented in detail in: Evangelou, P. "Analysis of Constraints to Expanded Livestock Production in Kenya's Maasailand." Ph.D. Thesis, University of Florida, 1984.

7
Livestock Marketing

This chapter focuses upon cattle and small stock marketing at the local level, with the objective of not only examining the extent to which marketing inefficiencies hinder the transition process in Maasailand, but also identifying constraints to the emergence of a more efficient marketing system. The interregional cattle trade, dominated by the supply of slaughter stock to urban markets, is considered first, followed by an examination of Maasailand's internal trade in cattle and small stock. The last section draws together conclusions regarding livestock marketing efficiency in Maasailand, and culminates in an assessment of the extent to which production is constrained by market operations. Discussion begins with a description of the theoretical approach taken in the analysis.

ASSESSING MARKET PERFORMANCE

The performance of a marketing system has two aspects, frequently referred to as technical efficiency and price efficiency [Aldington 1979; Purcell 1979]. Technical efficiency is attained when goods and services are provided at minimum average cost, that is, when the least-cost combination of marketing activities are employed. As the name suggests, technological improvements are usually the source of increased technical efficiency. Price efficiency refers to the capacity of the marketing system to adjust to changing supply and demand conditions. The market is considered relatively price efficient if there is a smooth flow of information along marketing channels and participants are able to readily modify their allocation of resources in response to price signals.

The practices of sampled livestock traders operating in the vicinity of the KGR and WNP sample areas and local-level butchers provide a basis for assessing market performance. Commercially oriented producers in Maasailand, such as the owners of the more progressive of the sampled individual ranches, usually do not utilize the common trading channels but, rather, rely upon on-ranch sales. Market mechanisms which are described pertain to the pastoral Maasai majority.

Analysis of the interregional cattle trade is approached in terms of the structure-conduct-performance (S-C-P) model originally developed in the study of industrial organization [Scherer 1970], in which market performance is attributed to the conduct of buyers and sellers in such matters as pricing and degree of cooperation or collusion. Conduct in turn is related by the model to the market's structure, that is, such characteristics as the number, size, and spatial distribution of buyers and sellers, and the relative presence or absence of barriers of entry into the trade. Recognition of the underlying influence upon both market structure and participant conduct of various basic conditions affecting supply and demand, from the availability of substitute products to laws, regulations, and dominant socioeconomic values, completes the S-C-P theoretical construct.

Operations of local-level butchers are then analyzed, with particular attention to the small stock trade and determinants of local-level supply and demand. Qualitative inferences regarding the interregional and intraregional trading networks are considered in terms of traders' and butchers' actual costs and returns. Analysis of price margins and operating expenses serves to substantiate conclusions with respect to market performance.

INTERREGIONAL CATTLE TRADING

Market Structure

The basic structure of cattle marketing is largely the same throughout Maasailand, with most initial transactions occurring at the producer's home. Livestock traders buy stock which are then frequently sold at a local market usually located in a town or trading center. The trade may follow one of several channels, depending upon the type of animal and the transactor's objectives, from the final sale locally of breeding stock to the intermediate sale of slaughter animals ultimately destined for a distant urban market. Exemplifying this structure are the patterns of trade found in the WNP and KGR sample areas. In the former area, two local-level markets are found at the towns of Kilgoris and Sakek (Fig. 7.1). Twice a week in each of these towns buyers and sellers gather in customary clearings, with the cattle offered for sale usually numbering 30 to 70 head. For the KGR sample area, the town of Emali is the site of the first formal market (Fig. 7.1). The Emali market, scheduled once a week and drawing from a larger area of supply than the Kilgoris and Sakek markets, averages about 280 head on offer.

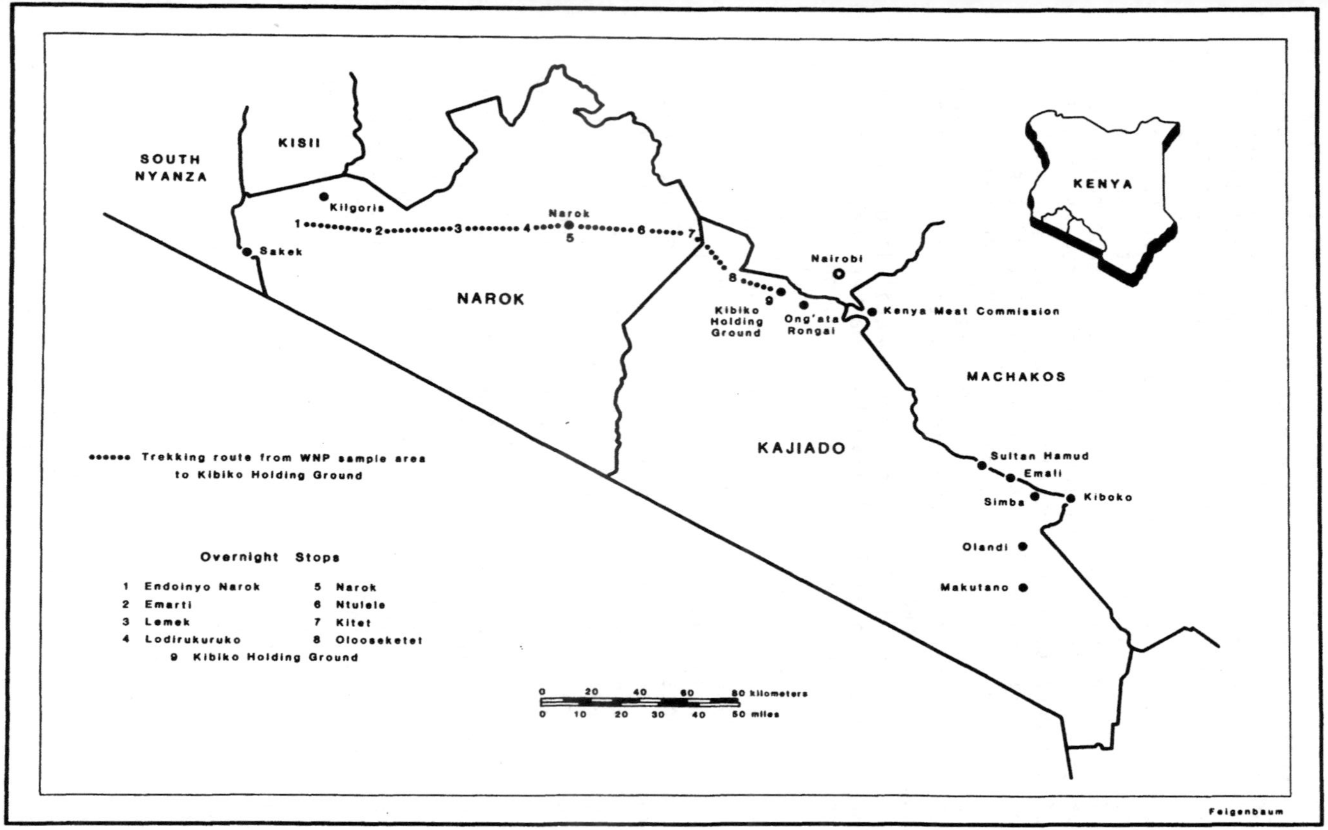

Figure 7.1. Livestock marketing samples

Though the markets in the two sample areas differ substantially in size, they are alike in their regulatory and infrastructural characteristics and more innately in the types of trading channels which they link. Livestock markets in Maasailand operate under the jurisdiction of local governments. For example, at the sample markets, district-level county council and subdistrict-level location officials validate interregional transactions by the collection of a cess, or sales tax, of Ksh 7 per head. Local transactions among Maasai are exempted from the cess. In addition, livestock taken from the sample areas, either to another district or markets within Maasailand which directly serve Nairobi, are inspected by officials of the Department of Veterinary Services. Movement permits are required in an effort to control the transmission of diseases, especially foot-and-mouth. Permits specify the number of animals, the destination, the number of days expected to be in transit, and the route to be taken.

There is little direct governmental involvement in Maasailand's local-level livestock markets beyond these revenue collection and disease control measures. The Livestock Marketing Division (LMD) is relatively inactive in this part of the country, in contrast to its quarantining activities in northern Kenya, and several of the holding grounds established in Maasailand by LMD have subsequently come under the authority of local governments. Infrastructural development of the sample markets ranges from minimal to nonexistent. The Kilgoris and Sakek marketplaces are no more than open fields. The Emali market, slightly better-equipped, has a corral where cattle are kept temporarily following sale and a chute used in the examination of cattle by veterinary officials.

More intrinsic to the market structure are the underlying trade patterns that prevail. Sakek and Emali straddle district (tribal) boundaries, a common feature of Maasailand's markets which underscores the importance of localized, interregional trading networks. The district boundary is particularly noteworthy at Emali, with the separation of the cattle market on the Kajiado side from the market for small stock, and heifers on occasion, on the Machakos side. This jurisdictional (cess collecting) division is understandable, since almost all of the cattle offered for sale originate in Kajiado District and the majority of small stock traded come from Machakos District.

The predominance of cattle over small stock in the Maasai livestock trade is exemplified by the proportions of species traded by 43 traders interviewed by the author at Emali (Table 7.1). Only two of them traded strictly in small stock, and one considered it merely a part-time occupation, trading "only when still waiting rain." Still, over one-third of the 41 cattle traders stated that they had started by trading in small stock, presumably because

Table 7.1. Livestock traded by sampled traders, Emali, Kajiado District, 1980

Species	Traders	
	Number	Proportion of Total
		--percent---
Cattle only	29	68
Cattle and small stock	12	27
Small stock only	2	5
Total	43	100

of the lower prices involved. Their particular transactions on the days of the interviews are summarized in Table 7.2. A total of 203 head were brought to market, for a mean number of five head per trader. Most of the cattle were Small East African Zebu, with only 11 percent (22 of the 203) of improved breed: 11 Boran-cross, 10 Sahiwal-cross, and one Friesian-cross. Seventy percent were purchased directly from producers, about one-fourth from fellow traders, and 6 percent had been part of the traders' home herds.

Neighboring and more densely populated non-Maasai districts are sources of breeding stock for the Maasai, as well as outlets for young Maasai-bred bulls and steers desired as draft animals. Feeder stock are traded in both directions, but with the major flow outward from Maasailand. However, as described in Chapter III, cattle sold by Maasai are principally slaughter stock destined for Nairobi and other urban markets. Thirty percent of the cattle brought to Emali by the traders referred to above were immatures sold for breeding and draft purposes, and 70 percent were adults marketed for slaughter. By sex, over 70 percent were steers, with the balance fairly evenly divided between bulls and females. The overwhelming majority of immatures were steers (Table 7.3).

These relative proportions of breeding/draft and slaughter sales are similar to percentages recorded for 2,584 head of cattle sold at Emali over a one-year period beginning September, 1981 (one-fourth of the estimated number of total transactions), as shown in Table 7.4. Sixty-two percent of the cattle were sold for slaughter, with the majority channeled to urban centers, while 38 percent were sold to producers in Maasailand and Machakos District for breeding or draft purposes. Distinct slaughter and breeding/draft demands determine the trade patterns for the WNP sample area as well.

Cattle marketing in Maasailand, then, can be visualized as a funneling of stock, usually by traders, from households to local markets at which local-destination transactions principally involving nonslaughter stock are overshadowed by the sale of slaughter stock destined for Kenya's urban markets. It is a flexible structure, involving differentiated products (by sex and age of stock), numerous buyers and sellers, and little direct regulation beyond veterinary inspection and the collection of a sales tax. The major regulatory influence is in fact indirect, namely, the impact of meat price controls discussed in Chapter III. Their pervasive effect on market performance is reconsidered in the last section of this chapter.

Trader Conduct

Livestock traders in Maasailand are generally young men in their twenties for whom "producer-trader" would be

Table 7.2. Cattle transactions of 41 sampled traders on the day of interview, Emali, Kajiado District, 1980

Feature	Purchases		Sales	
	Quantity	Proportion of Total	Quantity	Proportion of Total
	--head--	--percent-	--head--	--percent-
Purchased from/ Sold to another trader	48	24	96	47
Purchased from/ Sold to a producer	143	70	42	21
Homebred cattle	12	6	N.A.	N.A.
Cattle not sold	N.A.[a]	N.A.	15	7
Unknown: (i) whether or not sold, and (ii) if sold, whether to trader or producer	N.A.	N.A.	50	25
Total	203	100	203	100

[a]N.A. = not applicable.

Table 7.3. Sex and age characteristics of cattle brought to market by 41 traders on the day of interview, Emali, Kajiado District, 1980

Sex	Age	Quantity	Proportion of Total
		--head--	--percent--
Female	Adult	27	13.3
	Immature	5	2.5
Male	Adult	21	10.3
	Immature	4	2.0
Male castrate	Adult	94	46.3
	Immature	52	25.6
Total		203	100.0

Table 7.4. Destinations of cattle traded at Emali, Kajiado District, 1981/1982

Purpose	Destination	Number[a]	Proportion of Total
		-head-	--percent-
Slaughter	Ong'ata Rongai	510	20
	Other Nairobi Markets	214	8
	Kenya Meat Commission--Athi River	242	9
	Mombasa markets	220	9
	Emali	25	1
	Other	380	15
Subtotal		1,591	62
Breeding or draft	Machakos District	612	24
	Kajiado District	381	14
Subtotal		993	38
Total		2,584	100

Source: Adapted from Bekure, Evangelou, and Chabari [1982], Table 2.

[a]One-fourth of estimated total transactions, September, 1981, to September, 1982.

a more accurate name, since they invariably own nontraded herds and flocks. The indistinctiveness of the trader identity is compounded by the fact that transactions may be sporadic and take place solely within a local market's catchment area, comprise a full-time and interregional business, or constitute any temporal-spatial combination between these extremes. District authorities require livestock traders to be licensed, charging an annual fee averaging Ksh 100, but enforcement is lax. Number of transactions is also not a meaningful criterion by which to distinguish the Maasai trader from the Maasai producer. For example, the head of WNP 1 bought and sold over 100 head of cattle during 1981 for a cumulative gross margin (selling price minus buying price) of over Ksh 10,400, yet did not consider himself a trader.

The lack of uniformity among traders' operations is partly attributable to the ease of entry into and exit from the trade. There are no barriers to becoming a livestock trader, as long as one has the time and money (although it is unlikely that a successful career could be established by a non-Maasai person). Some traders find little incentive to alter a trading network once established, given the advantages of familiarity and routine. Others who are less restricted by home-herd responsibilities and have greater financial resources, chart wider-ranging and varying trading paths. Thus, one finds overlapping trading spheres, with traders drawing from pools of producers and frequenting markets in a variety of patterns. Invariably, however, primary concern among traders is with the personal herd. Even when livestock trading is a full-time occupation, it becomes secondary in importance when obligations concerning one's home herd arise.

Established trading spheres do not signify or imply sole rights to a purchasing area. On the contrary, traders emphatically acknowledge that a trader may buy livestock anywhere and from whomever has animals for sale, producer or fellow trader. Of the 2,981 head of cattle offered at the Emali market for which it was recorded whether the seller was a trader or producer (about one-fifth of the 13,500 head estimated offered for sale during the survey year beginning September, 1981), 96 percent were brought to market by traders [Bekure, Evangelou, and Chabari 1982]. Forty-one percent of these cattle had not been purchased directly from producers, but rather from other traders, a percentage even greater than that indicated in Table 7.2. Similarly for the Kilgoris and Sakek markets, the majority of market participants are traders, with pre-market transactions between traders equally prevalent.

Although overall a markedly individualistic trade, 13 of the 41 cattle traders interviewed at Emali described partnerships with one or sometimes two other persons. Usually partners buy cattle from producers in different trading areas, thereby enabling a larger producer region to be covered. In only one reported partnership did the traders often accompany each other while buying livestock. The purpose of partnerships is apparently to reduce income fluctuations. Earnings are shared, or more commonly, loaned back and forth between partners as needed. If one trader makes large profits at the market one week and his partner does not, funds can be loaned to the latter, allowing each one's transactions to continue unabated. However, such arrangements are not the norm, and perhaps only represent a formalization of the frequent credit agreements which characterize the trade. Traders operate independently in the buying and selling of livestock, despite a camaraderie and group identity even beyond that found generally among Maasai. Occasional partnerships, grounded in friendship, stabilize incomes while not impinging upon an individual partner's trading sphere.

Traders operate individually when buying stock, but following purchase there is a high degree of coordination in the actual movement of animals to markets. In the KGR sample area, traders will often group purchased cattle at customary sites and have them trekked by herdboys to Emali, with arrival timed for the day of the market. This coordination results in lower operating costs for the traders as well as reducing the risks associated with trekking. Three-fourths of the 41 Emali cattle traders interviewed employ herdboys, though a few indicated that they do so only occasionally. Similar arrangements characterize the movement of stock within the WNP sample area to a lesser extent. The trekking of stock to the Nairobi market from the WNP sample area, on the other hand, is invariably coordinated by groups of traders.

Price discovery, whether at a producer's home, at a market, or in transit, is by one-to-one negotiation. There are no auctions at markets, but rather numerous individual transactions taking place simultaneously on a willing buyer--willing seller basis. An animal may be sold more than once on the same market day, adding complexity to the trade. Given the individualized nature of the transactions, not to mention the absence of sales by liveweight, traders' profits are earned by the keen eye and bargaining acumen.

As mentioned, credit is used extensively in the trade, especially between traders but also between trader and producer. Frequently transactions involve payments which are either partially or wholly deferred until a later date. The fact that credit is so readily extended among traders reflects the personal friendships which permeate the trade, and suggests default is not common (although extensions beyond the agreed upon period for repayment may well be). Widespread use of credit allows on average more transactions to occur than would otherwise take place.

When asked why they trade livestock, most traders describe consumption demands, with food purchases most commonly mentioned. Second among reasons for trading is the opportunity, using the profits, to increase the size of the home herd. As noted by Doherty [1979b], trading can be a means of quickly acquiring a personal herd for a young man. But the attraction of trading is more than economic, as most traders view it as also simply a way of making life more enjoyable than it would be otherwise, "staying at home looking after cattle." Contributing to this positive aura, cattle trading is one of the few forms of regular employment which does not compromise the Maasai life-style. One's cultural identity is fully retained, and for young men even enhanced.

Market articulation, then, depends upon Maasai traders whose trading activities are competitively independent and yet which incorporate coodinating cost and risk reducing arrangements and the informal but frequent use of credit. There are no barriers to entry into the trade, as signified by the blurred distinction between producer and trader, and scope of operation is limited only by one's efforts, home-herd responsibilities, financial capability, and trading skills.

Costs and Returns

The KGR sample area. An equivocal picture of market performance is inferred from this overview of market structure and trader conduct. Conditions both favorable and unfavorable to market efficiency exist, as summarized in Table 7.5. Additional insight regarding performance levels can be gained by the analysis of prices which characterize the trade. For the KGR sample area, mean costs and returns for slaughter cattle purchased from producers and marketed at Emali and Ong'ata Rongai were calculated by Bekure, Evangelou, and Chabari [1982], for a sample of 152 weighed Small East African Zebu. This breed comprised over 99 percent of 3,051 transactions recorded at Emali for which breed was clearly identified. The sampled animals were purchased from producers at a mean price of Ksh 1,012 per head, and sold at Emali for Ksh 1,396, which resulted in a net income (return to the trader after deducting marketing expenses) of Ksh 319, or 30 percent of the original purchase price. The mean price at Ong'ata Rongai, Ksh 1,919, permitted a net income again of about 30 percent on the Emali purchase price. The percentage added value between purchase from producer and final sale for slaughter averaged nearly 90 percent. Prices and costs per kg indicated in Table 7.6 are based on a 260 kg animal throughout. Since some weight loss probably occurs during movement from point of initial purchase to point of final sale, the values per kg are only approximations.

The large trading margins indicated are supported in general by the recorded transactions of the 41 cattle traders interviewed at Emali. The highest gross margins (selling price minus buying price) were received for four

Table 7.5. Inferred cattle marketing performance in the KGR and WNP sample areas, 1981

Criterion	Existing Condition	Impact on Performance
Market Structure		
Number of buyers and sellers	Many	Favorable
Entry and exit	No barriers	Favorable
Flow of information	Mixed	Mixed
Regulatory controls		
Direct	Jurisdictional, disease control	Favorable
Indirect	Meat price controls	Unfavorable
Trader Conduct		
Price discovery	One-to-one negotiation	Unfavorable
Provision of credit[a]	Extensive	Favorable
Coordination of stock movement	Frequent	Favorable

[a]Informal credit arrangements among traders and between traders and primary producers.

Table 7.6. Mean costs and returns to cattle traders at Emali and Ong'ata Rongai, Kajiado District, 1981/1982

Item	Cost or Return	
	Per Head	Per kg Liveweight[a]
	----------Ksh----------	
Purchase price paid	1,012	3.89
Marketing costs to Emali		
Direct		
Trekking fee	20	
Watering fee	2	
Loss--Trading	10	
Death (1/60)	17	
Indirect		
Food and lodging	12	
Personal transport	4	
Total	65	.25
Sale price at Emali	1,396	5.37
Trader's net income at Emali	319	1.23
Purchase price paid at Emali	1,396	5.37
Marketing costs to Ong'ata Rongai		
Direct		
Cess	7	
Trekking fee	20	
Watering fee	2	
Loss--Trading	14	
Death (1/60)	32	
Indirect		
Food and lodging	20	
Personal transport	12	
Miscellaneous	12	
Total	119	.46
Sale price at Ong'ata Rongai	1,919	7.38
Trader's net income at Ong'ata Rongai	404	1.55

Source: Bekure, Evangelou, and Chabari [1982], Table 7. Adjusted for discrepancies.

[a]Assumed liveweight of 260 kg throughout.

adult steers in two separate transactions: Ksh 600 per head for two of them and Ksh 550 per head for the other two. Percentage gross margins (gross margin divided by buying price, multiplied by 100) for animals purchased during the week prior to the day of sale ranged from less than one percent to (on two occasions) 68 percent, with all three animals adult steers. In one instance, an immature Boran-crossed steer purchased one full year earlier for Ksh 470, was sold for Ksh 1,000, earning the trader a percentage gross margin of 112 percent. Excluding this steer, the mean percentage gross margin for the remaining 131 head of cattle was 25 percent (standard deviation, 16.6 percent).

Average rates of return of 30 percent to the trader's capital, management and personal labor over a mean period between purchase and sale of a week--even a percentage gross margin (that is, without considering operating expenses) of 25 percent--seem unexpectedly high, given the competitive appearance of the trade. Traders can exert a degree of bargaining power but probably not overly so, given producers' alternative marketing possibilities, from selling to competing traders to having a member of the household take the animal to market. However, producers' trading options may be limited and bargaining positions weakened by urgent cash demands. In some instances, the producer may attempt to time a sale in order to take advantage of seasonal fluctuations in demand, but as indicated in previous chapters, the decision to sell an animal is rarely foremost a function of expected price.

The large trading margins at Emali can be partially explained by the frequency of nonsales. Bekure, Evangelou, and Chabari [1982] estimate that one out of five head of cattle supplied to the Emali market is not sold. Table 7.7, which shows the responses of 40 Emali cattle traders when asked about the frequency of their nonsales, supports this estimation. Readily available post-nonsale options result in wider margins than would otherwise occur. The trader can return home with the stock (or arrange for someone else to care for them) and then offer them for sale again the following week, or simply carry them on to Ong'ata Rongai, bypassing the Emali transaction. Table 7.8 shows the actions taken by 29 traders subsequent to nonsales.

The WNP sample area. Data on the transactions of three traders operating in the WNP sample area permit additional insight into traders' costs and returns. The sampled traders, designated A, B, and C, follow distinct trading patterns. Trader A engages almost exclusively in the trade to Nairobi, while B and especially C also rely upon sales at the Sakek market, within the WNP sample area. Table 7.9, which summarizes their household and home-herd circumstances, shows that Trader A comes from a larger and more wealthy household than do Traders B and C and attained a higher level of formal education, factors

Table 7.7. Frequency of nonsales of cattle for 40 traders at Emali, Kajiado District, 1981

Number of Traders	Stated Frequency of Nonsales
14	Once a month
11	Twice a month
6	Not frequently
3	One-third of the time
2	Once or twice a month
2	Less than half of the time
1	More than half of the time
1	Frequently in times of drought

Table 7.8. Action taken by 29 traders subsequent to the nonsale of cattle at Emali, Kajiado District

Subsequent Action	Number of Traders
(i) Cattle are taken to the market at Ong'ata Rongai	2
(ii) Cattle are returned to the Emali market the following week	12
Either (i) or (ii), depending upon the age of the animal and other circumstances	15

Table 7.9. Background information on Traders A, B, and C, WNP sample area, Narok District, 1981/1982

Item	Trader		
	A	B	C
Age	27	23	26
Formal education	4 years secondary	None	Primary
Dependents			
Adults	4	5	2
Children	8	4	5
Home herd			
Cattle			
Female			
Adult	156	60	48
Immatures			
Calves			
Male	260	40	
Adults			16
Immatures			15
Calves			
Steers	114	15	16
Total	530[a]	115	151
Sheep	30[a]	5	8
Goats	40[a]	6	15

[a]Owned jointly by Trader A and his father.

which may partly explain Trader A's greater involvement in the Nairobi trade. Adult steers constitute two-thirds of the three traders' livestock sales, reflecting their engagement principally in the slaughter trade (Table 7.10). The mean buying and selling prices, gross margins, and percentage gross margins for the three traders are found in the first three columns of Table 7.11. As expected, greater returns are acquired at the Kibiko Holding Ground than from sales within the WNP sample area.

Comparing Tables 7.6 and 7.11 the mean price per head at Ong'ata Rongai is much higher than at the Kibiko Holding Ground. This discrepancy may be largely due to differences in the sizes of cattle slaughtered. An average carcass weight for cattle slaughtered in the WNP sample area during the survey period was found to be 85 kg (n=69), whereas in the KGR sample area it was 114 kg (n=154). Differences between traders' returns in the two sample areas are even more unexpected. Whereas traders' margins after expenses approach 30 percent between Emali and Ong'ata Rongai, returns before expenses between the WNP sample area and the Kibiko Holding Ground for Traders A, B, and C were 18.1, 21.0, and 13.1 percent, respectively. While recognizing the limited confidence, statistically, which can be attached to findings based upon such a small sample, the transactions of the three traders are considered representative.

The difference between the two sample areas in average returns to traders is especially surprising given the relative distances livestock are moved. Movement of stock from Emali to Ong'ata Rongai and other Nairobi markets takes about three days, whereas eight to nine days are required from the WNP sample area to the Kibiko Holding Ground (Fig. 7.1). In the latter case, cattle bought over a period of several weeks are accumulated until their number justify the trek. The trader must have the financial capability to postpone the sale of purchased cattle for up to several weeks. Hence, the financial requirements and physical risks for traders from the WNP sample area who engage in the Nairobi trade are greater than for their Kajiado counterparts.

The Narok traders' post-nonsale options are limited upon arrival at the Kibiko Holding Ground, which may partly explain their lower gross margins. They can pay to water and graze their stock at the holding ground for a nominal fee and hope for a more favorable seller's market in the near future. But, the uncertainty of future prices and the cost in trading time foregone make this an unattractive alternative, as suggested by the fact that sales at less than the purchase price do occur--2 out of 22 recorded sales for Trader B and 4 out of 56 for Trader C.

Clearly, profits can be sizable for traders operating in both sample areas but especially those in Kajiado. However, the frequency of nonsales and possible

Table 7.10. Sex and age characteristics of sampled cattle marketed by Traders A, B, and C, WNP sample area, Narok District

Sex	Age	Trader A	Trader B	Trader C	Trader Total	Proportion of Total
		-----head-------				---percent----
Female	Adult	3	9	33	45	25.4
	Immature	-	1	7	8	4.5
Male	Adult	2	-	-	2	1.1
	Immature	-	-	-	-	--
Male castrate	Adult	27	20	70	117	66.1
	Immature	-	-	5	5	2.9
Total		32	30	115	177	100.0

Table 7.11. Prices, gross margins, and percentage gross margins for Traders, A, B, and C, WNP sample area, Narok District, and one small-scale trader, Emali, Kajiado District, 1981/1982

Place of Sale, Item	WNP Sample Area Traders			Emali Trader
	A	B	C	
	-----------Ksh per head-----------			
Kibiko Holding Ground				
Number of traded cattle included in survey	32	22	56	--
(1) Mean buying price	1,335	1,207	1,198	--
(2) Mean selling price	1,576	1,460	1,355	--
(3) Gross margin (2-1)	241	253	157	--
Standard deviation of gross margin	104	186	126	--
Sakek				
Number of traded cattle included in survey	--	8	59	--
(1) Mean buying price	--	643	673	--
(2) Mean selling price	--	732	732	--
(3) Gross margin (2-1)	--	89	59	--
Standard deviation of gross margin	--	30	48	--
Emali				
Number of traded cattle included in survey	--	--	--	56
(1) Mean buying price	--	--	--	791
(2) Mean selling price	--	--	--	884
(3) Gross margin (2-1)	--	--	--	93
Standard deviation of gross margin	--	--	--	66

Table 7.11. continued

Place of Sale, Item	WNP Sample Area Traders			Emali Trader
	A	B	C	
	percentage			
Kibiko Holding Ground				
Percentage gross margin[a]	18.1	21.0	13.1	--
Standard deviation of percentage gross margin	8.5	16.7	14.7	--
Sakek				
Percentage gross margin[a]	--	13.8	8.8	--
Standard deviation of percentage gross margin	--	10.9	7.1	--
Emali				
Percentage gross margin[a]	--	--	--	11.8
Standard deviation of percentage gross margin	--	--	--	8.6

[a] Percentage gross margin = $\frac{(3)}{(1)} \times 100$.

post-nonsale options need to be included in a realistic evaluation of traders' margins and market performance. Net incomes for Kajiado traders as represented by the values in Table 7.6 may be on the high side, with the distribution of trader margins skewed by occasional, exceptionally large profits. Foremost, it should be kept in mind that traders' costs and returns, as is true for all aspects of their operations, vary widely. The last column of Table 7.11 shows the mean buying and selling prices, gross margin, and percentage gross margin for a trader in the Kajiado sample area who sells two to three cattle per week at Emali. It is evident that his margins do not approach those indicated in Table 7.6.

INTRAREGIONAL TRADING

As recounted, the major share of mature cattle sold by the Maasai are destined for the urban markets of Nairobi and, secondarily, Mombasa. Adequate slaughtering facilities and a sufficient demand in some of the larger towns of Kajiado and Narok Districts permit local sales of beef, but the intraregional livestock trade is primarily in sheep and goats. Because of their size, small stock essentially require no slaughtering facilities. They provide fresh meat in towns and trading centers where demand is low and cold storage not feasible. Data from five butchers operating in Maasailand are used to investigate the local trade in small stock and, to a lesser extent, cattle. Two butchers who deal only in small stock, located in the towns of Kiboko and Simba, and another butcher who slaughters cattle as well as small stock, in Sultan Hamud, represent local meat marketing operations in the vicinity of the KGR sample area. The remaining two butchers are found in Narok District, where one located in the town of Narok sells mutton almost exclusively, and the other, operating within the WNP sample area in the town of Kilgoris, slaughters only cattle. The locations of these five butchers are shown in Fig. 7.1, and each butcher is identified in the following discussion by the name of the town in which he is located.

Species and numbers slaughtered by the butchers depend largely upon sources of supply and size and nature of demand. An important characteristic can be noted before discussing the supply and demand conditions faced by each of the butchers, namely, the common practice of preparing meat for immediate consumption. Butcheries throughout Maasailand, and most of Kenya for that matter, operate small eating houses, often in association with a restaurant or hotel. Meat purchased is cooked according to the customer's preference, usually roasted or boiled as a stew. By preparing a cut of meat for consumption, butchers can legally charge Ksh 1.50 per kg above the controlled retail price of beef, mutton, or goat meat [Kenya, Republic of 1983]. The pervasiveness of these eating houses suggests that they contribute importantly to a butcher's income. All of the sampled butcheries included this activity in their operations, two of them on their own and three together with a neighboring restaurant/bar. Keeping in mind this common trait, particular circumstances which distinguish each of the sampled butcheries are now considered.

The Kiboko butcher is located in a small community and sells only goat meat, averaging two head slaughtered every three days. Kiboko is little more than a few shops on the Nairobi-Mombasa highway, and a large proportion of the butcher's customers are travelers. Sheep are not slaughtered by this butcher because of their lower demand (customers generally prefer goat meat over mutton), and lower supply (sheep tend to be retained by producers, especially the Maasai, for home consumption due to their higher fat content). Since Kiboko is near the district boundary, as many of the goats purchased come from Machakos District as from Kajiado. The Simba butcher, also located on the Nairobi-Mombasa highway but in a larger community than the Kiboko butcher, handles a slightly larger sales volume which includes sheep as well as goats. However, goat meat is again predominant.

The trade of the Kiboko and Simba butchers is small in scale compared to the operations of the Sultan Hamud butcher, who slaughters cattle as well as small stock. He is the only one of the five butchers who worked at more than one butchery during the survey period, which helps to explain the greater number of slaughters recorded. Whereas the sampled Kiboko and Simba butchers each have two to three competitors, in Sultan Hamud there are over a dozen butcheries. Sultan Hamud is a major town of the region, also situated along the Nairobi-Mombasa highway, but much more of an administrative and trading center than Kiboko or Simba. Community facilities enabling the slaughter of cattle, in contrast to the limb of a tree used by the Kiboko and Simba butchers for slaughtering small stock, make possible the sale of beef by Sultan Hamud butchers. Again, goats are slaughtered more frequently than sheep.

The operations of the Narok (Town) butcher present a different depiction of local-level meat marketing in Maasailand. Only small stock are slaughtered, but unlike the three Kajiado operations, sheep greatly outnumber goats. The explanation is again found in the sources of supply and the nature of the demand. The sheep are purchased weekly at Mulot, a market very near the border of Narok and Kericho Districts. The larger share of the sheep purchased by the sampled butcher are brought to this market by non-Maasai producers from more densely populated Kericho. Narok's interior location and importance as district headquarters result in a principally Maasai consumer base, resident and transient. Consequently, the sampled butcher has built up a sizable Maasai clientele by catering not only to the general demand for meat, but to their strong preference for the fat of sheep as well. Sheep fat is a major source of earnings, constituting from one-fourth to one-eighth of a sheep's carcass weight. Thus, the Narok butcher, located in a large town with many Maasai customers and with access to a major supply market drawing from non-Maasai regions, logically concentrates on the slaughter of sheep, averaging three head per day.

The last of the five surveyed butchers is found in the town of Kilgoris, divisional headquarters in the WNP sample area. Unlike the other four, this butcher sells only beef, using the community's slaughtering facilities about once every three days. The prevalence of cattle slaughters in the WNP sample area is thought to reflect a relatively smaller sheep and goat population compared to other parts of Maasailand, but livestock censuses are not available to verify this supposition.

The relative proportions of livestock purchased by the butchers directly from producers and from livestock traders varied. Small stock were fairly evenly supplied by producers and traders for the Simba and Sultan Hamud butchers. However, three-fourths of cattle purchases by the Sultan Hamud butcher were from traders, indicating the greater trader activity in cattle than sheep and goats discussed in the preceding section. Purchases from livestock traders outnumbered purchases from producers for the Kilgoris butcher as well, although the proportions were more balanced.

An estimation of the distribution by age of livestock slaughtered by the Sultan Hamud butcher is possible using recorded numbers of permanent incisor teeth (Table 7.12). Nearly all cattle were over 3 years, with the majority more than 3 1/2 years, while 70 percent of goats slaughtered were 3 years and older. The age distribution for females compared to that of males and male castrates included higher proportions of older animals, especially for goats.

Animals are not purchased by butchers on a liveweight basis, but rather on a willing buyer--willing seller basis like traders' transactions. Consequently, dressed weights must be relied upon when analyzing purchase prices per kg. Information collected on carcass weights and purchase prices for the surveyed butchers are summarized, by species and sex, in Table 7.13. There is no discernible pattern by sex for the slaughters recorded. The number of cows slaughtered by the Sultan Hamud butcher is half the total number of bulls and steers, whereas cows outnumbered bulls and steers for the Kilgoris butcher. The sex proportions for small stock cannot be meaningfully analyzed, due to the large percentage not classified. The price and weight means indicate that males and male castrates generally bring the higher prices and have the heavier carcasses for both cattle and small stock, although prices and weights range widely for each sex.

Table 7.12. Age and sex characteristics of cattle and goats slaughtered by the Sultan Hamud butcher, 1981

Species	Number of Permanent Incisor Teeth		Age and Sex Distributions 2	4	6	8	Total, by Sex	Proportion of Total
								--percent----
Cattle	Estimated age in months[a]		25.3 ± 2.3	32 ± 3	38.7 ± 3.7	Over 42.6		
	Sex	Female	--	--	11	15	26	32.9
		Male	--	--	9	14	23	29.1
		Male castrate	--	1	16	13	30	38.0
	Total, by age group		--	1	36	42	79	100.0
	Proportion of total		--	1.3	45.6	53.1	100.0	
Goats	Estimated age in months[b]		16 to 21	22 to 27	28 to 33	Over 33		
	Sex	Female	3	29	47	46	125	38.4
		Male	2	26	41	27	96	29.4
		Male castrate	1	24	47	33	105	32.2
	Total, by age group		6	79	135	106	326	100.0
	Proportion of total		1.9	24.2	41.4	32.5	100.0	

[a]From Semenye [1980].

[b]From Wilson, Peacock, and Sayers [1981].

Table 7.13. Price and weight characteristics of livestock slaughtered by surveyed butchers, 1981

Location of Butcher	Species	Sex	Number	Purchase Price			Carcass Dressed Weight			Carcass Price per kg
				High	Low	Mean	High	Low	Mean	
			-head--	---Ksh per head----			--------kg--------			-----Ksh-----
Kiboko	Goats	Female	74	160	50	105	20	5	8	13.13
		Male, male castrate	114	300	70	124	12	5	9	13.78
		N.C.[a]	60	138	60	93	10	5	7	13.29
Simba	Sheep	Female	27	180	70	118	14	6	10	11.80
		Male	8	225	100	162	16	9	14	11.57
		Male castrate	27	260	80	170	19	8	13	13.08
		N.C.	14	250	80	158	14	10	14	11.28
	Goats	Female	42	250	100	152	20	9	13	11.69
		Male	9	275	75	152	21	8	13	11.70
		Male castrate	53	310	80	177	24	9	14	12.64
		N.C.	43	290	80	152	24	9	13	11.69
Sultan Hamud	Cattle	Female	55	2,000	500	1,063	245	60	105	10.12
		Male	38	2,400	700	1,163	234	70	124	9.37
		Male castrate	61	1,950	500	1,142	200	60	115	9.93
	Sheep	Female	6	160	100	130	13	10	12	10.83
		Male	1	135	135	135	16	16	16	8.44
		Male castrate	4	180	80	142	17	11	14	10.14
		N.C.	21	400	70	161	33	8	15	10.73
	Goats	Female	134	340	60	147	30	7	14	10.50
		Male	109	400	70	170	35	7	18	9.44
		Male castrate	106	400	50	147	34	7	14	10.50
		N.C.	265	400	60	153	35	6	14	10.93
Narok	Sheep	N.C.	437	480	65	184	42	7	15	12.27
	Goats	N.C.	17	225	120	151	19	11	14	10.79
Kilgoris	Cattle	Female	39	1,100	450	767	124	40	77	9.96
		Male, male castrate	30	1,260	500	944	128	55	95	9.94

[a]Not classified by sex.

Carcass weights for the sampled butchers were higher than the national average for small stock and lower for cattle. Other than for the Kiboko butcher, small stock carcass weights averaged 14 kg compared to 12 kg for sheep and 11 kg for goats for Kenya overall [FAO 1982a]. On the other hand, the mean of the cattle carcass weights recorded by the Sultan Hamud and Kilgoris butchers was only 105 kg, compared with the national average of 130 kg [FAO 1982a].

The ability to determine on sight the break-even price for a live animal is quickly acquired, given price controls and the competitiveness of meat retailing. The price of mutton and goat meat at the time of the survey was Ksh 13.50 per kg and the price of beef with bone, Ksh 11.50 per kg, for the Simba, Narok, and Kilgoris butchers, and Ksh 14 and Ksh 12, respectively, for the Kiboko and Sultan Hamud butchers (located in Machakos District) [Kenya, Republic of 1979]. The final column of Table 7.13, listing the mean purchase price per kg carcass weight, reveals how closely butchers are able to estimate the returns from purchased livestock. A comparison of these purchase prices (per kg carcass) with the controlled retail prices affords an indication of the average gross returns per kg earned by the butchers. The operating expenses for the Narok butcher, itemized in Table 7.14, exemplify butchers' additional expenditures. The transport costs for sheep purchased weekly at the Mulot market and employees wages are types of expenses which vary from butcher to butcher. Other overhead costs, such as the payment of local cesses and the purchase of fuel for cooking meat, are similar for all operations. Butchers must also allow for income foregone due to the condemnation of diseased animal organs and even entire carcasses. All livestock slaughtered are supposed to be inspected by government health officials, a regulation more closely adhered to in larger towns than in the smaller communities. The frequency of condemnations recorded by three of the surveyed butchers, shown in Table 7.15, indicates that livers and lungs are organs frequently diseased.

All uncondemned parts of a slaughtered animal are marketed. The retail prices of cattle organs are controlled, but for small stock the prices are set by the individual butcher. Prices received by the Simba butcher for noncarcass sales are representative of those obtained by the other butchers (Table 7.16).

Each town has at least one drying shed where hides and skins are suspended and dried. Some butchers sell the hides and skins of slaughtered animals on a regular basis to tannery agents, while others store them, hoping for a price increase. During 1981, the price for a sheep skin, stretched and dried, averaged Ksh 3.50, and for a goat skin, Ksh 9.00. Cattle hides, sold by weight, averaged

Table 7.14. Direct operating costs of the Narok butcher, 1981

Item	Cost
	-Ksh per head of small stock-
County council cess, market of purchase	4.00
Transport to Narok[a]	6.00
County council cess, slaughterhouse[b]	5.50
Person to slit throat[c]	1.00
Person to skin the animal	1.00
Transport, from slaughter-house to butchery	3.00
Person to oversee the drying of the skin	2.00
Total	22.50

[a]By hired lorry. Price had risen in two years from Ksh 3 to Ksh 6.

[b]Ksh .50 of this total paid to the Health Inspector.

[c]In accordance with religious (Islamic) requirement.

Table 7.15. Condemnations, surveyed butchers, 1981

			Condemnations					
Location of Butcher[a]	Species	Observations	Liver	Lungs	Intestines	Pancreas	Heart	Entire Animal
		----head----	---------	---------	---------number---------	---------	---------	---------
Simba	Sheep	51	15	6	9	--	--	--
	Goats	88	20	6	4	--	--	--
Sultan Hamud	Cattle	154	19	19	1	4	1	6
	Sheep	11	4	1	--	--	--	--
	Goats	349	51	50	6	--	--	10
Narok	Sheep	272	143	--	--	--	--	--
	Goats	5	5	--	--	--	--	--

[a]Condemnations not recorded by Kiboko or Kilgoris butchers.

Ksh 5 per kg depending upon the condition. The prices of hides and skins can fluctuate dramatically depending largely upon European inventories, since 90 percent of Kenya's marketed hides and skins are exported, principally to Italy and Greece. Table 7.17 shows the prices paid by tanneries in Nairobi during August, 1981, for hides and skins. Prices were twice this level a few years previous to the survey, but prices have been low during the past five years due to a long period of oversupply in Europe. For example, one major Nairobi tannery had been storing purchased hides and skins since 1978.

At a more elemental marketing stage than the sampled butchers are small stock slaughters at minor trading centers in Maasailand. For example, at Olandi, a group of shops on the pipeline road near the Merueshi-Mbirikani boundary (Fig. 7.1), a sheep or goat is slaughtered two or three times a week, with purchased cuts roasted on the spot. Table 7.18 shows the prices paid for various parts of carcasses of goats slaughtered on three successive Wednesdays (market day) at another small trading center, Makutano, situated within Mbirikani group ranch (Fig. 7.1). Such small-scale retailing illustrates market operation at its most simple level, with customers buying meat by the piece or cut rather than by weight.

CONCLUSIONS

The distinctive feature of the development process is the expansion and evolution of markets [Johnston and Clark 1982]. The evidence presented suggests that technical and price inefficiencies, the two aspects of market performance defined at the beginning of the chapter, are not preventing expanded livestock production in Maasailand. In terms of the S-C-P model, a generally efficient level of performance is inferred from the market's structure and the conduct of traders and butchers. Though mean returns per sale for cattle traders, especially ones operating in the KGR sample area, may appear excessive, trading risks and post-nonsale options help to explain sizable trading margins. This positive assessment recognizes that marketing efficiency is a relative concept. As is true for marketing systems everywhere, narrow perspectives, habitual procedures, and institutionalized behavior restrain change and adjustment in the Maasai livestock trade.

It is apparent that the capacity of markets to become more efficient and thereby promote increased levels of production is limited because prices are prevented from coordinating activities, guiding production, and allocating resources, and because nonmarket oriented production practices continue to prevail. Improvements in market performance will be obtained only if there are concurrent changes in the circumstances surrounding the supply and demand for livestock: national pricing policies and

Table 7.16. Prices for noncarcass small stock items of sale, Simba butcher, 1981

Item	Price[a]
	---------Ksh---------
Liver	10.50 per kg
Intestines	6.10 per kg
Lungs, heart	3.00 to 4.00 per kg
Head	6.00 to 10.00 each
Legs	0.80 each

[a]Lungs, heart, head, and legs are boiled together, and therefore prices are for the items cooked. The soup is sold for Ksh .50 per cup.

Table 7.17. Prices for hides and skins, Nairobi, August, 1981

Grade	Cattle Hide	Goat Skin	Sheep Skin
	-Ksh per kg-	-Ksh per skin-	-Ksh per skin--
I	8.50	13.00	6.50
II	7.00	11.00	5.50
III	6.00	9.00	4.50
IV	4.00	5.00	2.00

Note: Prices are for suspension-dried, not ground-dried, hides and skins.

Table 7.18. Prices of parts of small stock carcasses sold on three market days, Makutano, Mbirikani group ranch, 1982

			Part of Carcass					
Date	Species	Sex	Head	Neck	Chest	Left Rib	Right Rib	Front Legs (pair)
			--------------	--------------	--------------Ksh	--------------	--------------	--------------
January 27, 1982								
	Goat,	male castrate	6	40	40	40	40	20
		male castrate	3	25	25	25	25	12
		female	5	30	30	30	30	14
February 3, 1982								
	Goat,	male castrate	5	35	35	35	35	24
		male castrate	8	35	35	35	35	20
		female	5	30	30	30	30	20
		female	5	30	30	30	30	14
	Sheep,	male castrate	8	15	15	15	15	10
February 10, 1982								
	Goat,	male castrate	5	20	20	20	20	14
		male castrate	8	35	35	35	35	16
		female	5	30	30	30	30	16
		female	5	30	30	30	30	14
		female	5	25	25	25	25	10
	Sheep,	male castrate	10	25	25	25	25	12

Table 7.18. continued

Date	Species	Sex	Part of Carcass							
			Hind Legs (pair)	Liver with Fat	Liver without Fat	Lungs and Pancreas	Intestines	Vertebrae Meat	Stomach	Total
			Ksh							
January 27, 1982										
	Goat,	male castrate	40	10	--	7	20	15	3	281
		male castrate	25	--	5	7	10	12	3.50	177.50
		female	30	6	--	5	15	10	2.50	207.50
February 3, 1982										
	Goat,	male castrate	30	--	6	7	15	12	5	244
		male castrate	30	--	5	7	15	12	--	237
		female	30	10	--	7	15	20	5	232
		female	24	--	5	7	15	10	4	204
	Sheep,	male castrate	20	--	4.50	7	10	12	3	134.50
February 10, 1982										
	Goat,	male castrate	20	--	5	8	14	12	5	163
		male castrate	30	--	8	7	12	15	6	242
		female	30	10	--	8	12	12	3.50	216.50
		female	24	--	6	8	10	10	5	202
		female	20	--	7	5	15	12	4	178
	Sheep,	male castrate	24	--	7	8	15	15	5	196

Source: Christie Peacock of ILCA's Kenya Country Programme.

Maasai producers' objectives. A measure of market effectiveness is the degree of stability and order achieved in pricing patterns, production, and growth over time. Both the individual system and society in general are hurt by highly variable production and price patterns [Parcell 1979]. Meat price controls tend to restrain price variability, but they can hardly be said to lead to a concurrent reduction in supply fluctuations when price is not the primary marketing motivation.

Prices must accurately reflect demand and producers must be responsive to the prices, in order for improved market performance to promote expanded production. This essential link between marketing and production remains underdeveloped in Maasailand, as illustrated by the following two examples. The use of vehicles for transporting cattle might be proposed as one way of improving market flow. At the time of the survey of traders, there was in fact an individual who was trucking cattle, albeit irregularly, from the WNP sample area to Nairobi markets. Renting a lorry, he would transport 14 to 16 head at a time from Emarti (one day's trek from the WNP sample area; Fig. 7.1) at a total cost of about Ksh 1,700. He would sell standard and commercial grade cattle on a carcass weight basis to the Kenya Meat Commission, while higher-grade stock were sold to private wholesalers and butchers. This individual even used vehicles to transport feeder cattle to private Kajiado ranches located near Nairobi.

This single transport operation can be expected to develop into a widespread marketing alternative only if prevailing production conditions change. As long as meat price controls prevent cattle/transport price ratios from rising to levels whereby the regular use of vehicles would improve service and reduce costs--improve the technical efficiency of the market--the commitment of vehicles to the full-time transport of cattle is unlikely to occur. At the producer end, with commercial objectives secondary at best and sale levels low and erratic, the regular supply of stock necessary for a major capital investment in transport vehicles remains uncertain. Hence, the circumstances support Sandford's observation that

> the advantages of lorries are likely mainly to arise from infrequent opportunistic access to lorry transport at moments of high demand (especially for shoats [sheep and goats]) or crisis, or of excess capacity in the general lorry-transport business of the sort requiring a highly decentralized decision-making and contractual organization, and not from the provision of permanent

> services by a specialist and centralized livestock-transporting organization. [1983, p. 213]

Economic comparisons of trekking and trucking have been documented for other areas of Subsaharan Africa. For example, as reported by Simpson and Farris [1982], Josserand and Sullivan's [1979] study of traders' margins in the 765-800 km movement of cattle from Upper Volta to Togo revealed a return of 12 percent on total investment for trekking and 8 percent for trucking. The emergence of trucking as the economically favored means of transport in Maasailand will depend upon appropriate changes in national pricing policies and producers' production objectives.

The second example of the dependence of improved market efficiency on production factors centers on the general lack of price information which characterizes the trade. It could be argued that individual negotiation of prices is inefficient, encouraging speculation and not yielding a level of market stability which could be attained by public auction, which would presumably reduce the costs of price discovery and streamline the flow of information. Yet the assumption that publicized prices would result in more efficient resource allocation by producers--increase the price efficiency of the market--is unfounded, given producers' nonmarket objectives. Market intelligence would be improved by the public auction of animals or even the purchase of livestock on a liveweight basis by local-level butchers, but the impact on production would depend upon producers' responsiveness to this additional information. Unused weighbridges installed by the LMD at various markets in Maasailand signify the futility of providing conditions for improved market performance if market participants do not perceive the benefits that could be derived.

The operations of sampled local-level butchers illustrate how the particular supply and demand conditions of a community or area shape the development of local retail meat sales. In Chapters V and VI, the larger proportion of small stock marketed by Olkarkar producers as compared with Mbirikani producers was noted as an important difference between production on the two group ranches. The Olkarkar producers are encouraged to sell small stock by the higher demand (consumption along the Nairobi-Mombasa highway) and compelled to do so by land pressures. Market involvement by the Maasai, of which small stock sales to local-level butchers is an important aspect, will grow as the demand for cash goods and services and production pressures lead to greater commercialized production. Notably, livestock purchases by the sampled butchers were largely from non-Maasai producers. Characteristics implicit in improved performance such as standardization and specialization of butchers' operations (purchases by weight, for example), are dependent upon market involvement of producers, given a competitive environment.

Local-level trading mechanisms in Maasailand are not hindering livestock development. Apparent sources of inefficiency--nonpublic price negotiations at poorly equipped markets, long distance trekking of stock, the absence of purchase by liveweight--do not create undue costs or constrain product flow, given present production conditions. Market efficiency will improve, but only in combination with (i) producers becoming increasingly market oriented and thereby providing a reliable supply, and (ii) national pricing policies accurately reflecting demand, the two conditions needed to induce positive change. Relatively efficient market performance in Maasailand implies that, in the balance between production and marketing advances, the developmental lead needs to be taken on the production side.

The Maasai have a long history of trading in their relationships with other peoples. Recent trade patterns involving Tanzania underscore the ability of Maasai traders to astutely appraise marketing alternatives, and take advantage of those perceived as most favorable [Salzman 1981b; Waller 1979]. Prior to and immediately following Kenya's independence (1963), the price paid for Maasai cattle by the Kenya Meat Commission was about one-half that in neighboring Tanganyika and the direction of market flow from Maasailand was southward. As Jacobs noted at the time:

> So aware are Kenya Masai of the higher prices paid in Tanganyika that there has flourished for years now an illegal market trade across the border, to which many Kajiado Administrators have deliberately turned a blind eye because it is an important source from which Masai can secure monies to pay their poll tax, as well as relieve overstocking, in the absence of adequate or fair marketing conditions in their district. [1963, p. 5].

The price ratio for livestock on the Kenyan and Tanzanian markets is now reversed and the direction of movement is northward, especially due to the relatively higher value of Kenya's currency. Jacobs [1978] estimates 60,000 to 100,000 head of cattle per year and at least as many small stock entered the Kenyan market illegally during the late 1970s. Traders are ready to respond when provided with incentives.

Despite such extensive trade patterns, the Maasai economy still largely operates by barter at the producer level, as demonstrated by the transactions of a hide-and-skins trader whose field of operation includes the KGR sample area. This trader's transactions over a six-week period in 1981 are recorded in Table 7.19, and especially significant are the many skin "purchases" which were in reality exchanges for grain and manufactured goods. The demand for consumer goods will provide the incentive for increased livestock sales, leading in turn to increased monetarization.

In other parts of Kenya, livestock are marketed under conditions significantly different from those found in Maasailand. For example, in Kenya's northern rangelands, disease control measures more stringently restrict stock movements and traders generally operate on a much larger scale. The sales over a one-year period for a typical trader operating in Samburu District are shown in Table 7.20. He buys cattle primarily at local markets, but will travel to producers' homes if they have 20 or more head for sale. The animals are collected at one of the LMD's holding grounds in the district, where they are vaccinated for foot-and-mouth disease and quarantined, during which time the trader is charged salt, dipping, and grazing fees. The cattle are then transported by rail to Ong'ata Rongai or other markets serving Nairobi, with the trader's total operating expenses, from purchase to sale, averaging Ksh 100 per head. The trade in small stock to Nairobi from this area is similarly in large quantities, with traders transporting 140 to 180 sheep and goats at a time using double-tiered lorries.

In comparison to traders operating in the quarantined north, the financial requirements of Maasai traders, even those operating from the WNP sample area, are not great. The Samburu District trader may have up to Ksh 300,000 invested in cattle destined for the Nairobi market at one time, with months separating sales. As he put it, to trade one has to be "financially strong." In the case of Maasailand, such trading concentrations do not exist. Improved market performance will depend not so much on changes in market structure or the trade mechanisms as on the emergence of favorable national pricing policies and increasingly market oriented production. Increased marketing efficiency, in turn, will then provide an incentive for expanded production.

Table 7.19. Transactions by a hides and skins trader in the vicinity of the KGR sample area over a six-week period, 1981

Date	Purchased	Sold	Price/Circumstances
Aug 8	1 Goat skin		Ksh 4
" 9	10 Goat Skins		Ksh 40
" 9	1 Sheep		In exchange for one bucket (Ksh 80).
" 10		Sheep purchased Aug 9	Ksh 120
" 10	120 Goat skins		Ksh 500, 100 skins from one person and 20 from another.
" 10	26 Sheep skins		Ksh 26
" 12		100 Sheep and goat skins	Ksh 1000, at Olandi to an agent.
" 12	100 Jerry cans		Ksh 800, Nairobi, transported to trading area at a cost of Ksh 50.
" 15		100 Goat skins (approx)	Ksh 2,600, sold on a weight basis.
" 15	100 Goat skins		Ksh 500
" 17	100 Sheep skins		Ksh 125
" 17	15 Goat skins		Ksh 75
" 18	30 Goat skins		Ksh 54
" 20	204 Goat skins		Two skins for one 2 kg packet of maizemeal purchased in Nairobi (Ksh 3.60 per packet, including transport to trading area).
" 21	20 Goat skins		Ksh 70

Table 7.19. continued

Date	Purchased	Sold	Price/Circumstances
Aug 21	50 Goat skins		In exchange for maizemeal, as above.
" 21	10 Sheep skins		In exchange for one cup (Ksh 7.50).
" 25	72 Goat skins		In exchange for maizemeal, one 2 kg packet for 3 skins.
" 25	100 Goat skins		Ksh 500, and six cups.
" 27	39 Sheep skins		In exchange for 11 cups (Ksh 1 each), 7 for 25 skins and 4 for 14 skins.
" 27	360 Sheep and goat skins		From another trader, in exchange for 6 cups, 14 plates, and 24 blankets.
" 27		14 bundles (24 kg each) of maizemeal	Ksh 781.80, in the trading area.
" 27	1 Goat		In exchange for 6 packets of maizemeal and one tin of cooking fat.
" 27	160 Goat skins		Ksh 707
" 28	40 Sheep skins		Ksh 80
" 28	25 Sacks of onions		Ksh 2,500, in trading area.
" 28	36 Goat skins		Ksh 93
" 28	120 Goat skins		In exchange for 120 cups (Ksh 5 each).
" 31		25 Sacks of onions	Ksh 3,125 (Ksh 125 per sack), Nairobi, no transport cost (Government vehicle).
Sept 2	2 Goat skins		In exchange for one cup (Ksh 5).
" 2	6 Goat skins		In exchange for three cups.

Table 7.19. continued

Date	Purchased	Sold	Price/Circumstances
Sept 2	8 Goat skins		In exchange for 4 cups.
" 5		300 Goat skins	Ksh 6,000, Nairobi.
" 5		207 Sheep skins	Ksh 1,025, Nairobi.
" 8	36 Goat skins		Ksh 10 and 6 cups.
" 8	35 Goat skins		Ksh 175
" 9	100 Goat skins		Ksh 500, purchased after brought across the Tanzanian border.
" 9	50 Sheep skins		Ksh 23
" 10	600 Goat skins		Ksh 3,000, from Tanzania.
" 10	100 Sheep skins		Ksh 50
" 11	100 Goat skins		Ksh 500
" 13		700 Goat skins	Ksh 8,000, Nairobi.
" 13		100 Sheep skins	Ksh 500, Nairobi.
" 15	200 Sheep skins		Ksh 100, from Tanzania.
" 15	300 Goat skins		Ksh 1,500, from Tanzania.
" 16	27 Goat skins		In exchange for 7 jerry cans (Ksh 25 each).
" 16	60 Goat skins		Ksh 150
" 16	250 Goat skins		Ksh 1,250
" 16	100 Sheep skins		Ksh 50
" 21	90 Goat skins		In exchange for 25 cups (Ksh 3 each).
" 23	206 Goat skins		Ksh 1,000, from Tanzania.
" 23	60 Sheep skins		Ksh 120, from Tanzania.

Table 7.19. continued

Date	Purchased	Sold	Price/Circumstances
Sept 24	1 Goat		In exchange for 7 jerry cans (10 liter, Ksh 18 each).
" 24	100 Goat skins		Ksh 500
" 25	8 Goat skins		Ksh 32
" 26	30 Goat skins		Ksh 90
" 26		7 Goats	Ksh 1,120, to a butcher near Sultan Hamud.
" 27		1,380 Goat skins	Ksh 26,000, Mombasa.
" 27		29 Cattle hides	Ksh 8,007.50 (Ksh 18 per kg), Mombasa.

Table 7.20. Cattle sales to the Nairobi market over a one-year period by a livestock trader purchasing in Samburu District

Date of Sale	Quantity	Mean Price Purchase	Mean Price Sale
	--head--	-----Ksh per head------	
August, 1980	200	800	1,200
November, 1980	250	1,000	1,320
March, 1981	320	1,020	1,320
July, 1981	249	1,030	1,345

8 Transition and Development

Maasai pastoralists do not derive their livelihood from livestock sales. As is true for pastoral peoples throughout Subsaharan Africa, the subsistence orientation is predominant, with food security and reduction of production risks more important than maximization of net market returns [Bohannan and Dalton 1971; Dyson-Hudson 1980c; Gryseels 1983; Quam 1978]. Yet, in the analyses which have been presented, it is evident that a climate of change currently pervades the pastoral Maasai economy. A movement from subsistence to commercial production, while far from uniform (as illustrated by the wide range in practices of the production units sampled), represents an advancing transitional frontier as producers perceive opportunities and incentives to modify existing management practices. This shift is not unique to the pastoral Maasai, but rather is a prevailing characteristic of African agriculture in general [Collinson 1980; Upton 1973].

The significance and desirability of Maasailand's transition for the future welfare of all Kenyans have been presented. It has been shown to be a process driven by the interplay between external circumstances, particularly increasing population pressures and producers' changing perceptions and objectives. The purpose of this chapter is to summarize conclusions regarding hindrances to the transition process and propose policy changes necessary for their removal, that is, to identify the overriding constraints preventing expanded livestock production in Maasailand, present development options, and recommend policy guidelines for overcoming the constraints. Secondarily, the effectiveness of the approach which has been taken in analyzing constraints to Maasailand's livestock development is evaluated.

It is well to restate here objectives set forth in Chapter I:

1. Identify and evaluate factors exogenous and endogenous to the production unit which constrain the transition in Maasailand from subsistence to commercially oriented livestock production. This objective entails:

a) consideration of the history of the transition process, in terms of principal components and their interactions,

b) examination of national policies and activities, in particular price controls, which influence the process,

c) investigation of the influence of regional economic, sociocultural and tenurial conditions upon the process, and

d) analysis of (i) producers' production and marketing practices and productivity levels, and (ii) the efficiency of local-level livestock marketing mechanisms.

2. Recommend policy changes which could lead to increased livestock production in Maasailand by facilitating the transition from subsistence to market oriented production.

The various parts of the first objective have been met in previous chapters. In this chapter, the second objective is addressed.

OVERRIDING CONSTRAINTS TO MAASAILAND'S LIVESTOCK DEVELOPMENT

Price Incentives and Resource Control

Maasai livestock production and marketing have been examined in terms of the interactions between physical, biological, and socioeconomic factors, in the context of producers' and national goals. The analyses lead to the conclusion that institutional changes are necessary before producers can be expected to respond to technological opportunities. Economic development depends ultimately upon underlying institutions and policies [Donovan 1983]. Simply put, expansion of livestock production in Kenya's Maasailand foremost requires unbiased pricing policies and resource use controls.

A price structure for Kenya's livestock industry which is not biased against producers stands first among needed institutional reforms. The responsiveness of Maasai producers to price incentives, as revealed in the studies cited in Chapter III, demonstrates the importance to the transition process of a price structure which realistically reflects supply and demand conditions. Moreover, Maasailand's local-level livestock trade was found in Chapter VII to be relatively efficient, given prevailing price controls and the generally noncommercial orientation of producers. Reluctance on the part of Maasai pastoralists to sell stock implies that policy changes are in order, particularly in the area of pricing [Galaty 1981a; Hjort 1981; Sandford 1982]. Schultz's observation regarding the role of prices in general in promoting agricultural development is reaffirmed in the case of the Maasai producer:

> Despite all that has been written to show that farmers in poor communities are subject to all manner of cultural restraints that make them unresponsive to normal economic incentives in accepting a new agricultural factor, studies of the observed lags in the acceptance of particular new agricultural factors show that these lags are explained satisfactorily by profitability. [1964, p. 164]

Control over grazing and watering resources is the second key institutional issue determining Maasailand's transition. Price incentives to increase production are sterile if the opportunity to respond to these incentives is absent. The brief history of group ranches indicates that they have only served to delay commercialization, the very process group ranch formation was intended to generate. The original group ranch concept was an ambitious plan whereby the law would be used as an instrument of social change [Coldham 1979]. A principal original function of each group ranch committee was to allocate and enforce grazing rights of individual members [Davis 1970; Jahnke et al. 1978]. As has been described, this essential duty has been widely ignored, and control at the ranch level over members' stocking rates and management practices generally does not exist. Instances of members evading even the most straightforward of ranch responsibilities, such as contributing equitably to the purchase of acaricides for a group ranch dip, have become commonplace. Initial concern with ensuring local support necessitiated excessive leniency in the implementation of management controls. However, resource use can no longer be left to each pastoral household's discretion if group ranches are to become profitable production entities [Hopcraft 1981]. This simple truth is at the root of the overriding resource use constraint to expanded livestock production.

The problem of resource use rights is obviated for the individual ranches yet, as shown by most of the ranches included in the KIR sample, privatization has not been a sufficient condition for improved resource utilization. On the individual as well as group ranches, it is evident that the transition process ultimately depends upon a shift in Maasai producers' decision-making frame of reference, a shift not easily accomplished since pastoral values, cognitions, and experiences often remain dominant components of their orientations [Mbithi 1977]. For instance, Sullivan et al. found Tanzanian Maasai to be motivated by subsistence objectives rather than commercial purposes when partaking of government-sponsored disease control measures; in their words, the "value of subsidized technology may be largely dissipated" without a set of policies for channeling to markets whatever impulse there is toward expanded output [1978, p. 158]. In sum, given the incentive of unbiased market prices and effective product demand, the transition process is a function of control over, and investment in, the resource base. The problem is one of establishing effective control where currently it does not exist and, where it does exist, facilitiating production augmenting investments. Adapting Norman's [1982] insight with regard to bringing about change in small farming systems, the necessary condition for overcoming constaints to livestock development in Maasailand (assuming an unbiased price structure) is resource control, while the sufficient condition is producers' willingness to take advantage of opportunities for expanding production given this control. As stated by Pratt:

> Unless, on the one hand, governments or agencies seeking to introduce range improvements or new ideas can deal with specific groups of people, with known livestock, living within a prescribed territory; and unless, on the other hand, the people are so organized that they can accept these inputs, and can develop internal responsibility and capability for the management of their land and its resources--then there can be little chance of success. [n.d., p. 1]

Transformation Versus Modification

An institutional approach to Maasailand's livestock development problems carries important implications for the formulation of effective governmental intervention. The position taken by the World Bank and others [Fumagalli 1977; IBRD 1982], is to attribute Africa's poor agricultural performance largely to a lack of technological improvements suitable for local conditions. Their analysis suggests that the goal should be change within the existing management system rather than its transformation. Planners are advised to avoid preconditions and prerequisites and, instead, concentrate on the "instrumental role of technology" [Anthony et al. 1979]. A similar philosophy of building upon slight modifications in existing practices to create the conditions for more fundamental change underlies the farming systems approach to development [Collinson 1982; Shaner, Philipp, and Schmehl 1982]. However, it is doubtful that livestock development interventions can be effective by only modifying existing conditions when resource control is the basic limiting factor.

Certainly, improvements are obtainable within the pastoral mode of production. As described in Chapter V, Semenye [1982] found significant differences in calf performance (rates of growth) by group ranch household. Wilson, Peacock, and Sayers [1981] observed a similar high level of variation in growth rates for small stock on Elangata Wuas group ranch. The herd with highest average weight had an advantage, at 18 months of age, of 6 kg for goats and 5 kg for sheep over the herd of lowest average weight. It was commented, "this is a reflection of individual management or herding abilities and emphasis on these coupled with selection for the fastest growing animals could lead to rapid improvement" [Wilson, Peacock, and Sayers 1981, p. 39]. Significant differences among households in levels of marketing activity have also been described. Bille and Anderson [1980] reported that two-thirds of small stock sales on Elangata Wuas group ranch were made by only 15 percent of the households surveyed. A wide range in sales has been likewise described in Chapter VI for the three producer samples. Pastoral production and marketing disparities such as these signify all the more the need to enforce resource sanctions in order that producers with better management capabilities are given the opportunity and incentive to excel.

Transformation rather than modification needs to be the development goal. Technical progress can advance little and investment planning becomes a futile exercise without an effective institutional framework [Little 1982b; Molnar and Clonts 1983; Sfeir-Younis and Bromley 1977], a fact obliquely acknowledged by practitioners of farming systems research in their reference to the essential role of infrastructural support and policy [Hildebrand and Waugh 1983]. Taking this perspective, breakthroughs in livestock technologies are of secondary importance to the transition. For example, disease control is fundamental to development, and producers will be willing to pay for it according to perceived benefits, but it does not affect the basic mode of production [Jahnke 1982; McCauley 1983]. There must be incentive for increased production (unbiased pricing policies) and conditions enabling producers to respond to this incentive. Secondary, micro-level interventions which influence specific management practices by facilitating capital investments or by modifying environmental and biological conditions will then find the institutional atmosphere conducive to success.

For the transition to occur, market production and market consumption will need to build one upon the other. In the words of Campbell, "the Maasai cannot be expected to reduce their livestock holdings until reliable and adequate alternative sources of food are available" [1979a, p. 16]. Even with less of a dietary dependence upon home-herd products, there remains the long-term viability issue, that is, the need to accumulate livestock as a cushion against ecological uncertainty. As noted by de Wilde [1980a] and discussed in Chapter IV, the supply of livestock to markets by pastoralists like the Maasai is largely determined by seasonal changes in range conditions and other nonprice factors. Before the rains, livestock are in poor condition and with the range productivity in decline, larger numbers are brought to market. After a rainy season has commenced, the range is revived, animals regain condition, and fewer are offered for sale.[1] Semenye [1980] concluded in his study of the cattle herds of Elangata Wuas group ranch that a higher offtake of steers and bulls was possible, but would not occur because of the need for extra animals as insurance against future losses. These marketing-production strategies exemplify the safety first decision-making criterion [Roy 1952]. Hence, recognition of a commercially oriented system as a source of both increased security and increased, stabilized incomes will govern the rate at which the transition occurs.

The importance of markets in not simply transmitting but actually generating these perceptions cannot be overemphasized [Kaldor 1979]. The transition process involves decisions which are complicated compromises, and the adoption of an increasingly commercial orientation will depend upon Maasai producers perceiving positive trade-offs [MacCrimmon 1973; Smelser 1976]. Marketing strategies, especially in the beginning, will be subject to increased uncertainty, and a flexible attitude to commercial production can be expected [Congleton 1978; Cyert, Dill, and March 1964; Lindblom 1964]. However, fundamental resource use controls will be necessary for even these first tentative shifts toward market production to occur.

Attempts to achieve development goals while maintaining the pastoral mode of production is an "unsolvable" problem [Schneider 1981b]. Expanded production will require resource control and subsequent resource investment [Bourgeot 1981; Low, Kemp, and Doran 1980; Maloiy and Heady 1965; Scoville and Sarhan 1978; Simpson 1970].

The Alternative

It is worthwhile to pause to consider the consequences of a policy which neglects the overriding institutional constraints, and attempts to continue along the current development path of passive acceptance of pastoralism's perpetuation. Under present circumstances, the expansion of livestock production will be negligible, since rates of offtake of cattle are as high as can be expected given the pastoral mode of production: ". . . careful studies of the Kaputiei section of the Masai have shown that even under unusually favorable conditions there is little real surplus under present forms of management" [Brown n.d., p. 4]; ". . .the Masai in Kajiado District have been selling virtually all the cattle that are saleable since the early 1970s. . ." [White and Meadows 1980b, p. 1]; ". . .on the available evidence there would seem to be no supply problems from Kajiado and Narok District--if the price is right, cattle owners will sell" [White and Meadows 1978, p. 3]. In fact, a decrease in offtake rates may occur as the population density increases, if subsistence objectives remain paramount.

Lack of resource controls will continue to impair the signals and incentives necessary to guide and induce the Maasai to use their land more efficiently. Productivity will decline, with the repercussions becoming more severe as demographic pressures increase. For Kajiado District and the more arid parts of Narok District, grazing capacities will be pressed and surpassed. Per capita production of meat and milk will fall as the ratio of livestock to people falls. Intergroup and intragroup conflict will increase, as resources become increasingly scarce while institutions for allocating and regulating their use remain ineffectual. Large numbers of displaced pastoralists, driven by necessity from an overburdened resource base, will seek jobs in towns and cities, adding to the nation's surplus of untrained unemployed.

The outlook is only less somber in the near term for regions of higher agricultural potential such as the WNP sample area. These areas can support a denser population and land use pressures will be felt less immediately, especially while barriers to non-Maasai in-migration stand. But population pressures cannot be avoided in the long run, as non-Maasai Kenyans are attracted by the area's cultivable potential [Harwitz 1978]. The regression from an unimproved to an overexploited resource base as population levels mount will occur wherever land use controls remain unimposed. Rather than expanding its livestock production, Maasailand will be hard pressed to maintain current offtake levels.

Pastoralism's increasing inadequacy will be especially evident during periods of environmental stress. The livestock losses during the drought of the mid-1970s and the large amounts of relief aid required pointedly demonstrate the futility of continued reliance upon the pastoral mode of production. In Campbell's [1976b] investigation of the drought's impact on Maasai pastoralists, agropastoralists, and non-Maasai farmers in the Loitokitok area of Kajiado, the decline in the sizes of cattle herds for the pastoral households ranged from 30 to over 50 percent of pre-drought numbers. At the time of the survey, in 1977, over 60 percent of the households were reported to have insufficient livestock to provide for their subsistence needs. The agropastoralists suffered less hardship because they could draw upon a wider range of productive resources, while the farmers were the group hardest hit. As population densities increase, the pastoral strategy for viability will be adequate for an ever smaller proportion of the population, while dependence solely upon rain-fed crops will provide the least security. In the telling phrase of one Maasai elder, "now there are always more children, but the land does not give birth."

Pastoral intervention in Kenya has become "something of an ideology which at once confers legitimacy and conserves the system, by acknowledging the need for change while promoting only marginal change" [Ake 1981, p. 156]. Only through institutional change leading to effective control and exploitation of resources will Maasailand's productive potential be realized, and the suffering of the past be prevented from recurring in the future.

EFFECTIVE PASTORAL INTERVENTION

A Selective Focus

A dynamic process of adaptation and response has been described as the essence of the pastoral household's traditional strategy to maintain viability. However, viability implies not only adaptation in marginal matters, but systemic transformation reaching deeply into the structure of the particular system when its survival is at stake [Deutsch 1977]. In Maasailand, the transition to market oriented production, including crop cultivation in the higher-potential areas, is the only alternative to declining levels of production and welfare. Effective pastoral intervention for facilitating the transition will require a prioritized, selective, yet comprehensive program: institutional adjustments will need to precede micro-level interventions, governmental involvement will need to build upon private sector initiative, and the macro-level impact of the transition process will need to be acknowledged and prepared for at the national level.

Methods of communal grazing control have been proposed for Maasailand and other pastoral lands, from setting stocking quotas or grazing fees, to cooperative grazing arrangements [Crotty 1980; Hopcraft 1981; Sullivan, Farris, and Simpson 1982]. Among the more novel is a proposal by Goldschmidt [1975] for regulating resource use by establishment of "livestock savings banks" in which pastoralists could make physical deposits and withdrawals of cattle and small stock. But effective control is the critical and absent factor. In the extreme, subdivision of the group ranches for private use, while considered by some to be economically preferred, is a politically unfeasible alternative [Jacobs 1978; Jarvis 1980; Lele 1975; Livingstone 1976]. Besides, the privatization of communal holdings does not guarantee increased production, as shown by the KIR sample ranches and as reported for other parts of Subsaharan Africa [Roe and Fortmann 1982]. Lawry concludes in the case of Botswana that the "problem is predominantly one of identifying and promoting institutions capable of better regulating communal range use. . .[but that] prospects for endogenous, local-level action to organize the sharing and management of delimited communal grazing territories are probably not very great" [1983, p. 62].

A similar statement would hold for Maasailand. Still, organizational structures for implementing controlled resource use, individual and group ranches, are in place, and the Kenyan Government needs to operate through them in order to capitalize upon producers' initiative, that is, draw upon private sector energies [Elliot Berg Associates 1982]. The sampled production units analyzed reaffirm that differences in producers' priorities lead to different management strategies, different decisions for resource allocation, and different production techniques [Collinson 1979]. Efforts to promote effective control will be best rewarded when they focus on those ranches, group and individual, managerially capable of (i) enforcing resource-use limits, (ii) assuming responsibility for loans, (iii) undertaking capital investments which will heighten productivity, and (iv) charting a commercially oriented development path.

Grazing and investment decisions cannot be imposed upon producers. Suggestions of mounting "a comprehensive campaign. . .to try to convince pastoralists to sell more animals" [Devres 1979, p. 77] hardly deserve serious consideration. Rather, the minority of producers able and willing to assume the responsibilities and risks associated with fundamental changes in orientation need to be the targets of governmental intervention. Culture-specific institutions for pastoralists as for other peoples are not static. They are constantly "up for test," and changes will be adopted when it is clearly in the producer's interest to do so [Eidheim and Wilson 1979; Lewis 1975]. Realistically, contentment with traditional practices and objectives and with accustomed life-styles will not simply disappear in the process. Instead, a subsistence orientation will persist, but for a progressively smaller share of the population [Galbraith 1979].

By way of comparison, Kenya's agricultural history has shown that small-scale farmers have been very responsive to opportunities for profitable innovation and commercial production. For a wide range of crops--tea, coffee, pyrethrum, dairy products, sugar cane--smallholdings have proven to be the superior production units in Kenya [IBRD 1981]. Pastoralists will exhibit an equivalent responsiveness once resource control and production incentives are realized. The Government of Kenya can best utilize its limited resources by facilitating development efforts of those producers who are able and willing to take the transitional lead. The proposed selective approach would take advantage of differences in the readiness of producers, with eventual successes among the leading producers sharpening the impulse toward transition among the rest. But it is not an approach in which the government would embark upon the creation of an artificial environment--"an island of infrastructure" or economic enclave. Rather, the transition to commercially oriented production is viewed as primarily relying upon producers' initiative, and their ability and willingness to self-impose controls over resource use. Following, development alternatives which are possible once there are effective controls are briefly discussed.

Micro-Level Options

Micro-level improvements will have a realistic basis for adoption once there is control over resource use and effective marketing incentives. The variety of orientations of sampled producers and the wide range in resource availability and use give substance to Sandford's caveat that planners "must avoid generalizations which try to apply the same solution to a wide variety of pastoral situations faced with different circumstances" [1983, p. 126]. Ecological conditions stand foremost among factors distinguishing producers' situations. Therefore, discussion of feasible development alternatives will be divided between an examination of options appropriate for the more arid Kajiado sample areas and others more ecologically suited to the western Narok sample area.

The Kajiado sample areas. Three characteristics of the KGR and KIR samples provide possible guidelines for development: the lack of significant differences between the incomes and levels of marketing activity of the group ranch producers and individual ranches, the positive relationship for all producers between sales levels and incomes, and the importance of incomes from steer-feeding operations. The transition to increased commercial production by pastoralists would be unlikely to progress very rapidly if it were to depend solely upon perceptions of the monetary net gains attainable by operating like individual ranches. Net incomes are essentially the same for group ranch households as for private ranchers. However, pastoralists are not operating in a static society fixed by tradition but, rather, are experiencing rapid change. In particular, land pressures are influencing marketing attitudes, as exemplified by the differences in sales between Olkarkar and Mbirikani producers. Higher levels of sales by group ranch households are directly related to higher incomes and for that reason producers may well learn to appreciate the income earning potential of short-term steer-feeding enterprises. However, the communal range issue cannot be avoided. Ultimately even a regular program of short-term grazing of AFC-financed steers will require control over resource use.

The KIR sample does not present a clear case in favor of private land tenure precisely because so few of the sampled ranches have taken advantage of the opportunities afforded by privatization. Measures of management skills such as calving rates, a key performance indicator in Jahnke's [1982] view, differ little for several of the individual ranches from those found on group ranches. Yet the concepts of improved management are well-known and applicable, as readily demonstrated over a decade ago by Skovlin [1971] in his analysis of a commercially oriented ranch in the vicinity of the Kajiado samples. Application of improved management techniques in tandem with phased investments are the missing ingredients necessary for expanded production.

The few success stories of the KIR sample indicate that high production levels are achievable. Within group ranches, advances are also possible. De Boer, Job, and Maundu [1982] budgeted an indigenous goat enterprise under traditional management, based on parameters from Elangata Wuas group ranch (Table 8.1). Annual net returns of Ksh 178 per AU were estimated, assuming a breeding herd of 100 does and three bucks, sale of 59 weaned kids (at 12 months), and culling of ten does and four surplus males. Even these returns could be easily exceeded with improved management and breeding.

The ecology of Kajiado suggests improved pastures are a possible development activity. The programming of varying seasonal feed supplies and varying requirements of different species of livestock would provide decision-making information on what could be a major area of resource investment [Brumby 1974]. Also, Laksesvela and Said [1978] observe that tropical grasses are generally of lower nutritional status than are temperate grasses; pasture improvement and/or ration supplementation are possible micro-level sources of increased productivity, once control over resource use becomes effective. The economic evaluation of dry-season forage plant introduction based on criteria of internal rate of return and availability of investment funds has been demonstrated for Botswana [ILCA/Botswana Production Unit (Botswana) 1978]. Investment in a hay-making enterprise as described by Sullivan et al. [1980] may be another feasible enterprise for reducing seasonal fluctuations in feed. Again, the possibilities for such an improvement in Maasailand reduce to a question of resource control, management, and availability of capital.

The WNP sample area. In considering micro-level development possibilities in this more humid environment, the shift to mixed farming operations cannot be overlooked. The subsistence cropping by the small-scale producers can be expected to evolve into increased production for the market, as exemplified by the

Table 8.1. Estimated costs and returns for East African goats, Elangata Wuas group ranch, Kajiado District, 1982

Item	Unit	Value
Production parameters		
Number of does	head	100
Number of bucks	head	3
Kidding rate	%	80
Doe replacement rate	%	10
Dipping (spraying) frequency	per month	1
Drenching frequency	per year	3
Liveweight for does	kg	31
Liveweight for bucks	kg	40
Liveweight for weaners for sale	kg	30
Mortality of matures	%	5
Mortality of kids	%	7
Sale age of weaners	months	12
Age at first joining	months	16
Prices liveweight		
Weaner kids	Ksh per kg	6
Cull does	Ksh per kg	6
Cull bucks	Ksh per kg	6
Average herd structure		
Does, one head equals 0.2 A.U.	A.U.	20.0
Kids, one head equals 0.1 A.U.	A.U.	7.4
Bucks and surplus males, one head equals 0.2 A.U.	A.U.	5.2
Replacement does, one head equals 0.2 A.U.	A.U.	2.0
Total	A.U.	34.6
Annual direct costs		
Drenching, at Ksh 9/head/year	Ksh	1,890
Dipping, at Ksh 6/head/year	Ksh	1,260
Replacement bucks	Ksh	0
Labor, 1 full-time at Ksh 250 per month	Ksh	3,000
Casual labor, at Ksh 9/month	Ksh	108
Minerals, salt, at Ksh 3/head	Ksh	630
Vaccinations, Ksh 1.15/adult	Ksh	156
Ksh 1.55/kid	Ksh	114
Total	Ksh	7,272
Annual revenues		
Weaner kids (59 x 30 kg x Ksh 6)	Ksh	10,620
Cull does (10 x 31 kg x Ksh 6)	Ksh	1,860
Cull surplus males (4 x 40 kg x Ksh 6)	Ksh	960
Total	Ksh	13,440
Annual net returns per herd	Ksh	6,168
Annual net returns per breeder	Ksh	62
Annual net returns per A.U.	Ksh	178

Source: DeBoer, Job, and Maundu [1982], Table 13.

Note: Budget is based largely on Wilson, R.T. "Livestock Production on Maasai Group Ranches. 1. A Preliminary Survey of the Goat and Sheep Populations at Elangata Wuas." East African Agr. For. Journal 43 (1978):193-199.

operations of the head of WNP 1. At the same time, opportunities for commercial livestock production by the larger, private producers are evident. As with the Kajiado samples, resource control and management are the principal areas of constraint.

Considering first small-scale producers, the role of individual initiative in expanding both livestock and crop production cannot be overestimated. The exceptional enterprise shown by the head of WNP 1 provides a model by which to measure production possibilities. This producer's recorded cattle transactions, from mid-January to mid-November, 1981, yielded a total income of Ksh 10,570. His average margin was Ksh 91 per head (range, Ksh 20 to Ksh 240; standard deviation, Ksh 49), with three-fourths of sales at the twice-weekly Kilgoris market, about 15 km from home. The only major breaks in his cattle trading activities during 1981 occurred at times of planting, harvesting, and marketing of crops, particularly in March-April and August-September.

Costs of crop production for small producers such as the head of WNP 1 are fairly stable, given their minimal use of purchased inputs. Selling prices, however, can fluctuate significantly, particularly by season. For instance, whereas a debe of maize sold at home brought Ksh 55 in January, 1981, by April of the same year the price had fallen to Ksh 20. Such price fluctuations may significantly influence producers' sale decisions and therefore the transition process. As another example, the second crop of onions for WNP 1 during the survey year was ready for the market in February, 1981. However, wholesale prices in Nairobi were so low, at Ksh 25 per net, Grade I, and Ksh 15 per net, Grade II, that he decided to store the onions until prices increased. By April, when the onions were finally marketed, the prices had risen to Ksh 55 and Ksh 45 for the two grades. Unfortunately, by then half of the crop had been lost, due to poor storage, and total returns barely covered production and transport costs. Thus, factors such as storage and transportation may hinder cropping developments more so than resource limitations, given that land and non-Maasai labor are readily available.

The relatively low population density of the WNP sample area and the excess labor supply available from neighboring districts will encourage agricultural expansion. Assuming one adult male consumes about two 90-kg bags of maize per year, and an average yield of hybrid maize ranges from 10 to 15 bags per ha, subsistence grain production is readily accomplished by the small-scale producers and increased production for the market would appear inevitable.

The cropping and commercial livestock transactions for WNP 2 and 3 more accurately represent prevailing orientations in the area, with crops grown almost entirely for subsistence and only five to ten head of cattle sold each year. Traditional pastoral orientations dominate activities on both small-scale units. The producers' objectives are to build up their herds by purchasing heifers, rather than concentrating on short-term steer-feeding operations, despite average milk offtake levels of one liter or less per cow per day.

The two larger private holdings sampled are in beginning stages of development, and for that reason it is instructive to consider the progress of an ongoing ranch similarly endowed ecologically. The cattle raising operations of a tea estate in Kericho District, where major changes have been initiated in a relatively brief span of time, are appropriate for comparison. The tea company owns a parcel of land of about 1,000 ha which lies within Eco-climatic Zone III, but which is unsuitable for growing tea. Therefore, a ranching operation was established. Boran cattle are bred following relatively simple management guildlines, which include:

(i) vaccination for foot-and-mouth and rinderpest twice a year by the Veterinary Department, vaccination for blackquarter and anthrax, dipping of stock once a week, deworming after weaning, and castrating and dehorning of bull calves at one month,

(ii) breeding heifers at 300 kg (26 to 30 months), and purchase of working bulls from stud farms, and

(iii) sale of steers above 400 kg (430 kg average) at the farm gate for Ksh 5.50 per kg liveweight, and sale of culled cows by the head at local markets.

Pasture improvement is the area in which management has had a major impact, as shown in Table 8.2. By 1981, 30 ha of the estate had been paddocked, with 13 ha of improved pasture. Of particular significance to the potential for replicating this development experience in the WNP sample area is the relative ease with which star grass was established. Once an area was paddocked all that was required was mowing, broadcasting seed, and application of sulfate of ammonia (21 percent nitrogen) at a rate of about 180 kg per ha. Moreover, fencing of pastures even without investing in their improvement yielded gains. Whereas steers reached sale weight in 36 months when not paddocked, only 33 months were needed when paddocked.

The advantages of improved pasture are apparent, but clearly there were also major costs involved, in particular capital investments in fencing and a watering system, and operating costs of pasture inputs. Also, it must be recognized that the gains achieved by the Boran likely would not be matched by the local zebu of the WNP sample area. Notably, the price at which steers were sold by the estate was only Ksh .50 per kg more than the average price at which the owner of WNP 5 purchased local zebu steers during the same period. Still, this example of relatively rapid development basically could be matched in the WNP sample area, given resource control, proper management capabilities, and investment funds.

Another possible avenue of development for the WNP

Table 8.2. Pasture improvement and rates of gain for steers, estate in Kericho District, 1975 to 1981

Period	Rate of Gain	Pasture/Other Improvements
	--kg per day--	
1975-1976	0.35	None
1977-1979	0.52	Bush clearing, bush control; establishment of watering points; grass establishment
1980-1981	0.85[a]	Paddocking; improvement of established grass

Source: Unpublished data.

[a]Sample of 70 steers.

sample area could be in expanded, upgraded small stock production. The plans of the owner of WNP 4 to establish a sheep operation were described in Chapter V. He is in the process of introducing improved breeds into the area. De Boer, Job, and Maundu [1982] have shown that the upgrading of sheep is a potentially major field for expanding production. In their budgeting of a Hampshire ram X Corriedale ewes enterprise (breeds intended to be introduced by the owner of WNP 4), annual net returns of Ksh 326 per AU were estimated. As with pasture improvements, upgrading of stock will require stricter managerial controls than generally exist at present.

Increased dairy production is yet another possibility. Wilemski [1975] estimates that for a grade cow in an environment similar to that of the WNP sample area to produce on average 2,000 kg of milk per year requires about 320 kg of concentrates. If the animal is only kept on pasture (0.6 ha), the average annual production is estimated to be about 1,300 kg of milk. Thus the costs of the 320 kg of concentrates (for example, 220 kg maize and 100 kg cotton seed cake) would need to be compared with the expected benefit of the additional 700 kg of milk. Even if the feeding were determined to be worthwhile, the greater risk of losses to disease of grade stock would need to be incorporated into the decision.

Areas of crop-animal linkage will become increasingly important to producers with the emergence and growth of commercial cropping [McGown, Haaland, and De Haan 1979]. Already livestock as an investment opportunity provide an incentive to increased crop production, oxen are used as draft animals, and crop residues provide fodder. The use of manure to enhance soil fertility and the sowing of fodder crops are other mutually supportive linkages possible. However, land use pressures may not yet be sufficient for drawing forth these inter-activities.

In the Kajiado sample areas and that of western Narok, opportunities for expanded livestock production can be readily identified. While cropping will necessarily become increasingly important in the latter region, the basic areas of constraint are the same as in drier parts of Maasailand: control over resource use, access to funds for investment, and improved management, with phasing of improvements--as exemplified by the development of the tea estate's cattle enterprise--essential to success. The proposed selective approach to pastoral intervention favoring units willing and able to confront these areas of constraint would best facilitate the transition. However, such a policy must also be socially and politically acceptable. This aspect of the proposed policy approach to development is now addressed.

Equity Issues

The success of a selective approach responsive to demonstrated readiness, as opposed to a program of indiscriminate technical intervention, will require that the approach be politically sound and pragmatically operable. The Government of Kenya faces an unhappy choice, but one in which the long-run benefits of a selective, yet comprehensive livestock development program in Maasailand clearly exceed short-run "benefits" of a crisis management approach [Simpson 1982b]. Short-term costs incurred by choosing the first option are neither politically nor economically insignificant. Foremost, meat price controls can no longer be allowed to favor urban consumers over rural producers. In the nutritional interest of the nation and especially of the poor, and given Kenya's low price elasticity and high income elasticity of demand for beef [Kivunja 1978], the government must concentrate its efforts on raising incomes and increasing output rather than administering meat prices.

Even more risky politically than freeing meat prices will be challenges of inequity and the potential for factionalism which can be expected to result from a policy which intentionally favors production units able and willing to impose resource use controls and thereby undertake investments necessary to the transition [Street and Sullivan 1983]. Opposition to a development strategy explicitly selective can be expected from critics who associate any development policy which is discriminatory with increased inequality, social conflict, and the impoverishment and eventual "proletarianization" of pastoralists [Cliffe 1977; Crotty 1980]. The growth of dualism is a genuine fear of African leaders, and strategies which promote the individual are often suspected of encouraging the economic exploitation of the rural majority [Collinson 1972]. Clearly, however, wealth inequalities among the Maasai are not solely, nor even principally, a derivative of the growing influence of the state and market in the region [Galaty 1981c], and concern with inequality needs to be tempered with attention to changes in absolute levels of living [Hazlewood 1978].

Chenery [1979] addresses the issue of measures of economic performance appropriate to transitional economies in which there may be a tradeoff between short-term growth and long-term restructuring of the economy. It is not an issue easily resolved. The disequilibrium associated with the transition in Maasailand, for example, needs to be considered in terms of the eventual production gains, especially since widespread poverty and large inequalities in income in Kenya are likely to continue [Ikiara and Killick 1981]. In the long run, the increased investment and production will lead to a growth in welfare. Economically stagnating conditions, not opportunities for expanded production, are the real source of growing inequalities.

The inequity issue is nonetheless significant, not least in reminding the government that livestock development in Maasailand is a national problem. There will be national consequences regardless of whether interventions take a selective and responsive form. Hence, producer assistance will need to be matched by programs which facilitate the training of surplus populations and which provide compensatory mechanisms to replace the traditional systems of social security [Njoka 1979]. Simply stated, scarce funds and yet scarcer human capital must be allocated (i) for greatest long-run net returns in livestock production and, (ii) in a training and relocation program in order that the costs of displacement may be minimized. Preparing individuals for occupations in non-Maasai areas and nonlivestock sectors of the national economy so that the rangelands can be developed within their capacity will need to be an integral part of the development strategy, particularly for Maasailand's more arid areas [Jahnke 1979, 1982]. Otherwise, dryland farming will inevitably expand in areas unsuitable for cultivation, as populations increase and the productivity of labor falls relative to the productivity of the land [Swift 1982].

SUMMARY AND GENERAL APPLICABILITY OF THE APPROACH

An approach to the study of contraints to production in pastoral areas of Subsaharan Africa has been demonstrated, based essentially on a systems perspective in which the examination of components at the producer level is augmented by the analysis of marketing and institutional factors. Determining an effective role for governments in planning pastoral interventions, in this case the Kenyan Government's interventions in Maasailand, has been the focus. Conclusions have been discussed at some length. In summary, all of the forms of pastoral intervention described in Chapter II and categorized by Goldschmidt [1981b]--altering of the pastoral environment, improving the livestock, providing marketing facilities, and changing pastoral institutions and values--have been attempted in Maasailand. Even the less pessimistic assessments of past interventions convey equivocality, referring to "second-best alternatives" [Jarvis 1980, p. 606], "mixed effects" [Konandreas and Anderson 1980, p. 1], or, at best, "positive returns to schemes which failed to attain their immediate objectives" [Ruthenberg 1980a, p. 28]. Evaluations in this vein can be expected of future micro-level livestock interventions in Maasailand if necessary adjustments in the pricing and tenurial frameworks in which producers and traders operate are not made.

Ayre-Smith [1976], writing about processes involved in increasing the efficiency of livestock production, identifies three types of activity: the removal of impediments, the provision of services, and the use of innovations. In Maasailand, impediments have been reduced, if not removed, and services have been provided, from the development of water resources to range and veterinary inputs. Innovations have also been implemented, particularly in the area of tenurial change. The critical distinction between interventions which innovate and those which are impediment-removing or service-providing is the behavioral change demanded of producers by innovations. The receptiveness of the Maasai to developmental change invariably has been inversely related to the social and economic adjustments thereby required. But the transiton to a market oriented production system is necessary and inevitable. In the words of Rimmer:

> Exchange is riskier than subsistence, especially for the inexperienced. Enterprise is more difficult to administer than routine. Competition does hurt established interests. Yet a lesson of history seems unequivocally to be that the avoidance of these risks and inconveniences is done at the expense of economic progress. A stationary economy may be secure, but it is not one in which nutrition is improving and life expectancy increasing. [1983, p. 160]

The Maasai continue to be described as resisting change [Hampson 1975]. But throughout the history of interventions in Maasailand, producers have welcomed technological development, including veterinary services, and have shown an admirable fund-raising record to pay for these interventions [Jacobs 1963, 1978; Spencer 1983]. Sullivan [1979], for example, found the Maasai pastoralists included in his study of livestock systems in Tanzania to be highly receptive to interventions for improving their livestock. It is when interventions make demands upon social structure, and threaten existing political and economic relationships, that reluctance to change is encountered, and rational decision making at the household level is mistakenly aggregated by outside observers into system-wide irrationality [Peacock, de Leeuw, and King 1982]. Even when change finally can no longer be formally avoided, accompanying socioeconomic demands will be informally evaded however possible. Group ranches have failed to facilitate intended changes in livestock production precisely because their success has depended upon the imposition of controls over resource use which are socially unacceptable to the pastoral Maasai.

Twenty years have passed since Jacobs' [1963] proposal for instituting resource controls, referred to in Chapter IV, and still they remain missing. Some sources view the transition he portrayed from cooperative to privately titled lands as inevitable, and consider present group ranches as no more than a temporary phase of development [Chemonics International Consulting Division 1977; Heyer 1976a]. In the opinion of others the subdivision of group ranches would create economically unviable units, destroying what opportunity there is for the development of progressive ranches [White and Meadows 1981]. Whichever point of view prevails, the transition of Maasailand to increasingly commercialized production, given an unbiased price structure, rests upon the legitimization and acceptance of development liabilities and managerial control. Fuguitt [1983] concluded from her study of property rights and agricultural efficiency that the adoption of a private property rights system in marginally cultivable areas of Kenya has not provided effective production incentives; she identified lack of capital and infertile land as the binding constraints. Yet, resource control is imperative if capital and land investments are to be forthcoming.

The selective approach to Maasailand's development which has been proposed should not be made to sound more simple or certain of success than it would be. Planning and implementing of any type of intervention involves processes fraught with uncertainty, which is all the more reason that goals must be well-defined before setting a development strategy [Simpson 1983]. If, as Ominde [1977] perceives for rural development in general, the principal aim of Maasailand's livestock development is the reduction of poverty and a more rational use of resources--including the adoption of new technologies--a selective and responsive approach is the only feasible course of action.

Eventually, some form of private tenure is likely to replace communal tenure, as the costs associated with communal rights rise and the derived benefits fall. Anderson and Hill [1977], in their examination of the transition from communal to individual tenure rights in the American West, found that a comparison of the benefits and costs of defining and enforcing property rights largely explains the evolution of private property institutions.

> As long as the benefits of eliminating the commons are low relative to the costs, there is little incentive for individuals to define and enforce private property rights; the tragedy of the commons is small. However, as the ratio of perceived benefits and costs changes, so will the level of definition and enforcement activity. [Anderson and Hill 1977, p. 213]

In many pastoral areas like Maasailand, this changing perception of costs and benefits is occurring, as land rather than livestock becomes the dominant resource defining socioeconomic relationships [Galaty 1980d]. Meanwhile the Kenyan Government, in the interest of both their pastoral and nonpastoral populations, will do well to facilitate this change with price incentives, a selective and responsive approach in support of producers' transition to commercially oriented production, and the development of a program to facilitate relocation. The potential Maasailand holds for expanded livestock production is too valuable to the Maasai and national economies to be left underdeveloped because of inadequate pricing policies, ineffectual resource use controls, and unprioritized intervention.

This set of conclusions has been derived from an analysis of conditions at several levels. As systems of production, Maasailand's group ranch households and individual producers are foremost human systems [Bennett 1978]. Economic change is not only possible but probable as national and regional linkages evolve, and the opportunity set widens. In response, producers' objective functions are undergoing continual modification as well. Only by employing an approach that has included examination of both the producers and their institutional fields of operation have overriding constraints to Maasailand's livestock development become evident.

Procedures such as Swift's [1981] rapid appraisal technique and the farming systems research method of determining recommendation domains--production units with similar management priorities, practices and problems for which a common set of development activities may be proposed [Byerlee et al. 1980]--can be effective approaches in the planning of livestock development projects in Subsaharan Africa. Bartenge [1980], for example, has divided Kajiado District into four recommendation domains based on an ecological division (whether the more difficult period for producers followed the short rains or the long rains) and a hierarchical division (whether livestock were of improved or unimproved breed). A classification such as Bartenge's helps in the categorization of principal production problems. But as is evident in the case of Maasailand, technical problems may well be of secondary importance in relation to the institutional issues of price and property. The demonstrated approach shows the advantage of examining obstacles to economic development in pastoral areas at both macro and micro levels, since neither focus by itself can reveal the dynamics of a pastoral livestock system or constraints to its development sufficiently [Jahnke et al. 1978]. Planning of interventions for Maasailand's development will inevitably be a social process derived from the clash of interests [Kornai 1979]. By considering the continuum of constraints as a whole, there is a greater probability that chosen strategies will have a realistic and effective policy basis.

NOTE

[1] This supply pattern is especially evident at times of weather-induced short-term meat shortages. Following a dry spell when grazing conditions change for the better, the supply of livestock invaribly falls. For example, on April 5, 1981, the headlines of Kenya's Sunday Nation read: "Nairobi Hit by Meat Shortages." The fall-off in supply, which was "expected to become acute," was attributed to "current rains following the harsh drought." In outlying towns in pastoral areas, meat shortages are frequently experienced at the onset of a rainy season (examples were reported in Kenya's Daily Nation on April 7, 1981, and November 6, 1981, and in The Standard on October 29, 1981). This situation implies that during periods of the year when grazing conditions are below average, markets depend upon a supply of animals which, were it not for the reduced availability of forage, would not be offered for sale.

Bibliography

Ake, C. A Political Economy of Africa. Harlow, UK: Longman House, 1981.

Aldington, T.J. "The Monitoring of Performance in Agriculture Markets and Its Control." Presented at a Workshop held at the Institute for Development Studies of the University of Nairobi, March 26-29, 1979. Printed in Price and Marketing Controls in Kenya, ed. J.T. Mukui, pp. 146-57. Institute for Development Studies, Occasional Paper No. 32, 1979.

Allan, W. The African Husbandman. New York: Barnes and Noble, 1965.

Almagor, U. "Pastoral Identity and Reluctance to Change: The Mbanderu of Ngamiland." Land Reform in the Making: Tradition, Public Policy and Ideology in Botswana, ed. R. Werbner. Special Number of the Journal of African Law 24 (Spring 1980):35-61.

Ambrose, J. "Land Use and Land Management Problems in Kenya." Development Planning in Kenya, ed. T. Pinfold and G. Norcliffe, pp. 101-17. Downsview, Canada: York University, Geographical Monographs No. 9, 1980.

Anderson, F.M., and J.C.M. Trail. "Inicial Application of Modelling Techniques in Livestock Production Systems under Semi-Arid Conditions in Africa." Presented at the Fourth World Conference on Animal Production, Buenos Aires, 1978. Printed in Livestock Productivity and Trypanotolerance, ILCA (Kenya) Working Document 22, 1981.

Anderson, T., and P. Hill. "From Free Grass to Fences: Transforming the Commons of the American West." Managing the Commons, ed. G. Gardin and J. Baden, pp. 200-16. San Francisco: Freeman and Co., 1977.

Anthony, K., B. Johnston, W. Jones, and V. Uchendu. Agricultural Change in Tropical Africa. Ithaca: Cornell University Press, 1979.

Aronson, D. "Must Nomads Settle? Some Notes Toward Policy on the Future of Pastoralism." When Nomads Settle, ed. P. Salzman, pp. 173-84. New York: Praeger Publishers, 1980.

Aronson, D. "Development for Nomadic Pastoralists: Who Benefits?" Proceedings, Conference on the Future of Pastoral Peoples, Nairobi, Kenya, August 4-8, 1980, ed. J. Galaty, D. Aronson, P. Salzman, A. Chouinard, pp. 42-50. Ottawa: International Development Research Centre, 1981.

Auriol, P. "Intensive Feeding Systems for Beef Production in Developing Countries." World Animal Review 9 (1974): 18-23.

Awogbade, M. "Fulani Pastoralism and the Problems of the Nigerian Veterinary Service." African Affairs 78 (1979):493-506.

Ayre-Smith, R. "Planning Beef Production in Developing Countries." Beef Cattle Production in Developing Countries, Proceedings of the Conference held in Edinburgh, September 1-6, 1974, ed. A.J. Smith, pp. 445-62. Edinburgh: University of Edinburgh, Centre for Tropical Veterinary Medicine, 1976.

Ayuko, L. "Ranch Organizational Structures in the Implementation of Pastoral Range and Ranch Development Projects." Paper presented at the Workshop sponsored by ILCA on The Design and Implementation of Pastoral Development Projects for Tropical Africa, Addis Ababa, February 25-29, 1980. Addis Ababa, Ethiopia: ILCA (Headquarters) Working Document 4, pp. 163-85, 1980.

Baker, P.R. "'Development' and the Pastoral People of Karamoja, North-Eastern Uganda. An Example of the Treatment of Symptoms." Pastoralism in Tropical Africa, Studies presented and discussed at the XIIIth International African Seminar, Niamey, December, 1972, ed. T. Monod, pp. 187-202. London: Oxford University Press, 1975.

Baker, P.R. "The Social Importance of Cattle in Africa and the Influence of Social Attitudes on Beef Production." Beef Cattle Production in Developing Countries, Proceedings of the Conference held in Edinburgh, September 1-6, 1974, ed. A.J. Smith, pp. 360-74. Edinburgh: University of Edinburgh, Centre for Tropical Veterinary Medicine, 1976.

Bartenge, J.K. "A Report on the Zoning of Kajiado District into Recommendation Domains to Facilitate the Planning of Appropriate Adaptive Research Programmes." Prepared for the Ministry of Livestock Development (Kenya), 1980.

Barth, F. "Capital, Investment and the Social Structure of a Pastoral Nomad Group in South Persia." Capital, Saving and Credit in Peasant Societies, ed. R. Firth and B.S. Yamey, pp. 69-81. London: George Allen and Unwin Ltd., 1964.

Bates, R. Markets and States in Tropical Africa. Berkeley: University of California Press, 1981a.

Bates, R. "Food Policy in Africa: Political Causes and Social Effects." Food Policy 6 (August 1981b):147-57.

Baxter, P.T.W. "Some Consequences of Sedentization for Social Relationships." Pastoralism in Tropical Africa, Studies presented and discussed at the XIIIth International African Seminar, Niamey, December, 1972, ed. T. Monod, pp. 206-25. London: Oxford University Press, 1975.

Behnke, R. "Production Rationales: The Commercialization of Subsistence Pastoralism." Prepared for the Ministry of Agriculture (Botswana), 1982.

Behnke, R. "Fenced and Open-Range Ranching: The Commercialization of Pastoral Land and Livestock in Africa." Livestock Development in Subsaharan Africa: Constraints, Prospects, Policy, ed. J.R. Simpson and P. Evangelou, pp. 261-84. Boulder, Colorado: Westview Press, 1984.

Bekure, S., P. Evangelou, and F.N. Chabari. "Livestock Marketing in Eastern Kajiado, Kenya (Draft)." ILCA (Kenya) Working Document 26, Nairobi, 1982.

Bennett, J. "A Rational-Choice Model of Agricultural Resource Utilization and Conservation." Social and Technological Management in Dry Lands, ed. N. Gonzalez, pp. 151-85. Boulder, Colorado: Westview Press, 1978.

Bernard, F., and D. Thom. "Population Pressure and Human Carrying Capacity in Selected Locations of Machakos and Kitui Districts." Journal of Developing Areas 15 (1981): 381-406.

Bienen, H. "The Impact of Colonialism on Modern Economic Patterns of Distribution." Imperialism, Colonialism, and Hunger: East and Central Africa, ed. R. Rotberg, pp. 225-49. Lexington, Massachusetts: D.C. Heath and Company, 1983.

Bille, J-C. "Natural Environmental Constraints in Livestock Development Projects." Paper presented at the Workshop sponsored by ILCA on The Design and Implementation of Pastoral Development Projects for Tropical Africa, Addis Ababa, February 25-29, 1980. Addis Ababa, Ethiopia: ILCA (Headquarters) Working Document 4, pp. 59-71, 1980.

Bille, J-C., and F.M. Anderson. "Observation from Elangata Wuas Group Ranch in Kajiado District (Revised)." ILCA (Kenya) Working Document 20, Nairobi, 1980.

Bille, J-C., and H. Heemstra. "An Illustrated Introduction to the Rainfall Patterns of Kenya." ILCA (Kenya) Working Document 12, Nairobi, 1979.

Bishop, A.H. "Letter to the Editor." World Animal Review 14 (1975):43.

Bohannan, P. "'Land,' 'Tenure' and Land Tenure." African Agrarian Systems, ed. D. Biebuyck, pp. 101-11. London: Oxford University Press, 1963.

Bohannan, P., and G. Dalton. "Markets in Africa: Introduction." Economic Anthropology and Development, ed. G. Dalton, pp. 143-66. New York: Basic Books, Inc., 1971.

Bonte, P. "Ecological and Economic Factors in the Determination of Pastoral Specialisation." Journal of Asian and African Studies 16 (1981):33-49.

Boulding, K. Economics as a Science. New York: McGraw-Hill, 1970.

Boulding, K. "Commons and Community: The Idea of a Public." Managing the Commons, ed. G. Hardin and J. Baden, pp. 280-94. San Francisco: Freeman and Co., 1977.

Bourgeot, A. "Nomadic Pastoral Society and the Market." Journal of Asian and African Studies 16 (1981):116-27.

Bram, R.A. "Tick-Borne Livestock Diseases and Their Vectors. 1. The Global Problem." World Animal Review 16 (1975):1-5.

Brandstrom, P., J. Hultin, and J. Lindstrom. Aspects of Agro-Pastoralism in East Africa. Uppsala, Sweden: The Scandinavian Institute of African Studies, Research Report No. 51, 1979.

Brown, G. "Agricultural Pricing Policies in Developing Countries." Distortions in Agricultural Incentives, ed. T. Schultz. Bloomington: Indiana University Press, 1978. Excerpts reprinted in Pricing Policy and Development Management, ed. G. Meier. Baltimore: The Johns Hopkins University Press, 1983.

Brown, L. "The Development of Semi-Arid Areas of Kenya." Paper prepared for the Ministry of Agriculture (Kenya), 1963.

Brown, L. "A Preliminary Report on the Need for an International Range Research Station in East or Central Africa." Prepared as part of a Ford Foundation Special Consultancy, n.d.

Brown, M. Farm Budgets: From Farm Income Analysis to Agricultural Project Analysis. Baltimore: The Johns Hopkins University Press, 1979.

Brumby, P.J. "Tropical Pasture Improvement and Livestock Production." World Animal Review 9 (1974):13-17.

Buck, N., D. Light, L. Lethola, T. Rennie, M. Mlambo, and B. Muke. "Beef Cattle Breeding Systems in Botswana: The Use of Indigenous Breeds." World Animal Review 43 (1982):12-16.

Byerlee, D., M. Collinson, R. Perrin, D. Winkelmann, S. Biggs, E. Moscardi, J. Martinez, L. Harrington, and A. Benjamin. "Planning Technologies Appropriate to Farmers: Concepts and Procedures." CIMMYT, El Batan, Mexico, 1980.

Callow, L.L. "Ticks and Tick-Borne Diseases as a Barrier to the Introduction of Exotic Cattle to the Tropics." World Animal Review 28 (1978):20-25.

Campbell, D.J. "Land Use Competition at the Margins of the Rangelands: A Proposal for Research in Kajiado District." University of Nairobi, Institute for Development Studies, Working Paper No. 299, 1977.

Campbell, D.J. "Development or Decline: Resources, Land Use and Population Growth in Kajiado District." University of Nairobi, Institute for Development Studies, Working Paper No. 352, 1979a.

Campbell, D.J. "Response to Drought in Maasailand: Pastoralists and Farmers of the Loitokitok Area, Kajiado District." University of Nairobi, Institute for Development Studies, Discussion Paper No. 267, 1979b.

Casebeer, R., and R. Denny. "Wildlife Competition with Cattle in Kenya." Prepared for the FAO (Rome) Range Management Special Fund Project 238, 1967.

Chemonics International Consulting Division. "Livestock and Meat Development Study." Prepared for the Ministry of Agriculture (Kenya). Washington, D.C.: Author, 1977.

Chenery, H. "Comparative Advantage and Development Policy." American Economic Review 51 (March 1961):18-51.

Chenery, H. Structural Change and Development Policy. Oxford: Oxford University Press, 1979.

Christensen, E. "Tick Control Project, Kericho District. Report from the First Year of Operation, 1.8.1979-31.7.1980." Ministry of Agriculture, Nairobi, 1980.

Ciriacy-Wantrup, S. Resource Conservation: Economics and Policies, rev. ed. Berkeley: University of California Press, 1963.

Cliffe, L. "Rural Class Formation in East Africa." Journal of Peasant Studies 4 (January 1977):195-224.

Cohen, J. "Land Tenure and Rural Development in Africa." Agricultural Development in Africa: Issues of Public Policy, ed. R. Bates and M. Lofchie, pp. 349-400. New York: Praeger Publishers, 1980.

Coldham, S. "Land Tenure Reform in Kenya: The Limits of Law." Journal of Modern African Studies 17 (December 1979):615-27.

Collinson, M. Farm Management in Peasant Agriculture. New York: Praeger Publishers, 1972.

Collinson, M. "Understanding Small Farmers." Presented at a Conference on Rapid Rural Appraisal, University of Nairobi, Institute for Development Studies, December 4-7, 1979.

Collinson, M. "Some Notes on the Farmer as the Client for Research." Presented at a CIMMYT Workshop on Methodological Issues Facing Social Scientists in Applied Crop and Farming Systems Research, Mexico, 1980.

Collinson, M. "Farming Systems Research in Eastern Africa: The Experience of CIMMYT and Some National Agricultural Research Services, 1976-81." Michigan State University, International Development Paper No. 3, 1982.

Commoner, B. "Summary of the Conference: On the Meaning of Ecological Failures in International Development." The Careless Technology: Ecology and International Development, The Record of the Conference on the Ecological Aspects of International Development, Washington University, Warrenton, Virginia, December 8-11, 1968, ed. M. Farvar and J. Milton, pp. xxi-xxix. Garden City, New York: The Natural History Press, 1972.

Congleton, R. "The Role of Information in Choice: Toward an Economic Theory of Knowledge and Decision Making." Ph.D. Thesis, Virginia Polytechnic Institute and State University, 1978.

Cossins, N. "Report on Group Ranch Administration and Finance." Prepared for the IBRD/International Finance Corporation, 1980.

Cronin, A. "Report on the Kenya Second Livestock Development Project." Ministry of Overseas Development, Manpower Planning Unit, London, 1978.

Crotty, R. Cattle, Economics and Development. Slough, UK: Commonwealth Agricultural Brueaux, 1980.

Cruz de Carvalho, E. "'Traditional' and 'Modern' Patterns of Cattle Raising in Southwestern Angola: A Critical Evaluation of Change from Pastoralism to Ranching." Journal of Developing Areas 8 (1974):199-226.

Cyert, R., W. Dill, and J. March. "The Role of Expectations in Business Decision Making." The Making of Decisions, ed. W. Gore and J. Dyson, pp. 288-314. New York: Free Press of Glencoe, 1964.

Dahl, G. "Ecology and Equality: The Boran Case." Pastoral Production and Society, Proceedings of the International Meeting on Nomadic Pastoralism, Paris, December 1-3, 1976, pp. 261-81. Cambridge: Cambridge University Press, 1979.

Dahl, G. "Production in Pastoral Societies." Proceedings, Conference on the Future of Pastoral Peoples, Nairobi, Kenya, August 4-8, 1980, ed. J. Galaty, D. Aronson, P. Salzman, and A. Chouinard, pp. 200-08. Ottawa: International Development Research Centre, 1981.

Dahl, G., and A. Hjort. Having Herds. Stockholm: Liber Tryck, 1976.

Dalton, G. "Economic Development and Economic Anthropology." Economic Anthropology and Development, ed. G. Dalton, pp. 348-60. New York: Basic Books, Inc., 1971.

Dasmann, R., J. Milton, and P. Freeman. Ecological Principles for Economic Development. London: John Wiley & Sons, Ltd., 1973.

Davis, R. "Prospects for Joint Production of Livestock and Wildlife on East African Rangeland: The Case of Kenya." Dar es Salaam, Tanzania: University College, Bureau of Resource Assessment and Land Use Planning, Research Paper No. 4, 1968.

Davis, R. "Some Issues in the Evolution, Organization and Operation of Group Ranches in Kenya." University of Nairobi, Institute for Development Studies, Discussion Paper No. 93, 1970.

Day, R. "Rational Choice and Economic Behavior." Theory and Decision 1 (1971):229-51.

De Boer, A., M. Job, and G. Maundu. "The Relative Profitability of Meat Goats, Angora Goats, Sheep and Cattle in Four Agro-Economic Zones of Kenya." Paper presented at the Third International Conference on Goat Production and Disease, Tucson, Arizona, January 10-15, 1982.

de Leeuw, P.N. "Aspects of Climate and Rangeland Productivity in the ILCA Study Area Kajiado District." Prepared for the ILCA Programme Committee, Nairobi, 1982.

de Leeuw, P.N., and C. Peacock. "The Productivity of Small Ruminants in the Maasai Pastoral System, Kajiado District, Kenya." ILCA (Kenya) Working Document 25, Nairobi, 1982.

de Montgolfier-Kouevi, C., and A. Vlavonou. "Trends and Prospects for Livestock and Crop Production in Tropical Africa." Addis Ababa, Ethiopia: ILCA (Headquarters) Working Document 5, 1981.

de Souza, M. "Food Production and the Household Economy." ILCA (Kenya) Internal Paper, Nairobi, 1980.

Deutsch, K. "Introduction." Ecosocial Systems and Ecopolitics: A Reader on Human and Social Implications of Environmental Management in Developing Countries, ed. K. Keutsch, pp. 11-20. Paris: Unesco, 1977.

de Vos, A. "Game as Food: A Report on Its Significance in Africa and Latin America." Unasylva 29, no. 116 (1977): 2-12.

de Vos, A. "Must Africa Suffer the Environmental Consequences of Tsetse-Fly Control?" Unasylva 30, no. 121 (1978):18-24.

Devres. Evaluation of the Kenya National Range and Ranch Development Project, U.S.A.I.D. Number 615-0157. Washington, D.C.: Devres, 1979.

de Wilde, J. "Price Incentives and African Agricultural Development." Agricultural Development in Africa: Issues of Public Policy, ed. R. Bates and M. Lofchie, pp. 46-66. New York: Praeger Publishers, 1980a.

de Wilde, J. "Case Studies: Kenya, Tanzania, and Ghana." Agricultural Development in Africa: Issues of Public Policy, ed. R. Bates and M. Lofchie, pp. 113-69. New York: Praeger Publishers, 1980b.

Doherty, D. "A Preliminary Report on Group Ranching in Narok District." University of Nairobi, Institute for Development Studies, Working Paper No. 350, 1979a.

Doherty, D. "Factors Inhibiting Economic Development on Rotian Olmakongo Group Ranch." University of Nairobi, Institute for Development Studies, Working Paper No. 356, 1979b.

Dolan, T., and A.S. Young. "An Approach to the Economic Assessment of East Coast Fever in Kenya." Current Topics in Veterinary Medicine 14 (1981):412-15.

Donovan, D. "Measuring Economic Performance." Finance and Development 20 (June 1983):2-5.

Doran, M., A. Low, and R. Kemp. "Cattle as a Store of Wealth in Swaziland: Implications for Livestock Development and Overgrazing in Eastern and Southern Africa." American Journal of Agricultural Economics 61 (1979): 41-7.

Dorner, P. Land Reform and Economic Development. Kingsport, Tennessee: Kingsport Press, 1972.

Drummond, R.O. "Tick-Borne Livestock Diseases and Their Vectors. 4. Chemical Control of Ticks." World Animal Review 19 (1976):28-33.

Duncanson, G.R. "The Kenya National Artificial Insemination Service." World Animal Review 16 (1975):37-42.

Dyson-Hudson, N. "Range Livestock Production Systems in East Africa, with Special Emphasis on Kenya." Prepared for the ILCA Long Term Plan, Addis Ababa, 1980a.

Dyson-Hudson, N. "The Development Experience." Prepared for the ILCA Long Term Plan, Addis Ababa, 1980b.

Dyson-Hudson, N. "Human Factors in the Design and Implementation of Livestock Development Projects in Africa." Paper presented at the Workshop sponsored by ILCA on The Design and Implementation of Pastoral Development Projects for Tropical Africa, Addis Ababa, February 25-29, 1980. Addis Ababa, Ethiopia: ILCA (Headquarters) Working Document 4, pp. 87-102, 1980c.

Dyson-Hudson, N. "Monitoring: Some Notes on Concepts and a Programme." Paper presented at the Workshop sponsored by ILCA on The Design and Implementation of Pastoral Development Projects for Troipcal Africa, Addis Ababa, February 25-29, 1980. Addis Ababa, Ethiopia: ILCA (Headquarters) Working Document 4, pp. 201-09, 1980d.

Dyson-Hudson, N., and R. Dyson-Hudson. "The Structure of East African Herds and the Future of East African Herders." Development and Change 13 (April 1982): 213-38.

Eidheim, H., and R.T. Wilson. "Status of the Baseline Study of Elangata Wuas and Kilonito Group Ranches." ILCA (Kenya), 1979.

Elliot Berg Associates. Encouraging the Private Sector in Somalia, U.S.A.I.D. Number AFR-0135-C-00-2045-00. Washington, D.C.: Author, 1982.

Ellis, P., and M. Hugh-Jones. "Disease as a Limiting Factor to Beef Production in Developing Countries." Beef Cattle Production in Developing Countries, Proceedings of the Conference held in Edinburgh, September 1-6, 1974, ed. A.J. Smith, pp. 105-16. Edinburgh: University of Edinburgh, Centre for Tropical Veterinary Medicine, 1976.

FAO. FAO Production Yearbook. Rome: FAO Statistics Series No. 40, vol. 31, 1978.

FAO. FAO Production Yearbook. Rome: FAO Statistics Series No. 40, vol. 32, 1979.

FAO. FAO Production Yearbook. Rome: FAO Statistics Series No. 40, vol. 33, 1980.

FAO. FAO Production Yearbook. Rome: FAO Statistics Series No. 40, vol. 35, 1982a.

FAO. The State of Food and Agriculture, 1981. Rome: FAO Agriculture Series No. 14, 1982b.

FAO/IBRD. The Outlook for Meat Production and Trade in the Near East and East Africa, in Two Volumes. Volume II: Livestock Development Country Studies. Washington, D.C.: Authors, 1977.

Fenn, M. Marketing Livestock and Meat, 2nd ed. Rome: FAO Marketing Guide No. 3, 1977.

Ferguson, D. "A Conceptual Framework for the Evaluation of Livestock Production Development Projects and Programs in Sub-Saharan West Africa." Center for Research on Economic Development, University of Michigan, n.d.

Firey, W. Man, Mind and Land. Glencoe, Illinois: Free Press, 1960.

Fitzhugh, H.A., and E.K. Byington. "Systems Approach to Animal Agriculture." World Animal Review 27 (1978): 2-6.

Found, W. "Decentralised Planning in Kenya, 1963-1975." Development Planning in Kenya, ed. T. Pinfold and G. Norcliffe, pp. 81-94. Downsview, Canada: York University, Geographical Monographs No. 9, 1980.

Frantz, C. "Contraction and Expansion in Nigerian Bovine Pastoralism." Pastoralism in Tropical Africa, Studies presented and discussed at the XIIIth International African Seminar, Niamey, December 1972, ed. T. Monod, pp. 338-53. London: Oxford University Press, 1975a.

Frantz, C. Pastoral Societies, Stratification, and National Integration in Africa. Uppsala, Sweden: The Scandinavian Institute of African Studies, Research Report No. 30, 1975b.

Frantz, C. "Fulbe Continuity and Change under Five Flags Atop West Africa." Journal of Asian and African Studies 16 (1981):89-115.

Fuguitt, D. "Property Rights, Economic Organization and the Development of the Agricultural Sector--Tanzania and Kenya." Ph.D. Thesis, Rice University, 1983.

Fumagalli, C. "An Evaluation of Development Projects among East African Pastoralists." African Studies Review, 21, no. 3 (1978):49-64.

Gaile, G. "Processes Affecting the Spatial Pattern of Rural-Urban Development in Kenya." African Studies Review, 19, no. 3 (December 1976):1-16.

Galaty, J. "The Maasai Group-Ranch: Politics and Development in an African Pastoral Society." When Nomads Settle, ed. P. Salzman, pp. 157-72. New York: Praeger Publishers, 1980.

Galaty, J. "Organizations for Pastoral Development: Contexts of Causality, Change, and Assessment." Proceedings, Conference on the Future of Pastoral Peoples, Nairobi, Kenya, August 4-8, 1980, ed. J. Galaty, D. Aronson, P. Salzman, and A. Chouinard, pp. 284-91. Ottawa: International Development Research Centre, 1981a.

Galaty, J. "Maasai Farmers, Pastoral Ideology, and Social Change." Presented at the Symposium on Planned and Unplanned Change in Nomadic and Pastoral Societies, Intercongress of the International Union of Anthropological and Ethnological Sciences, Amsterdam, April 23-25, 1981b.

Galaty, J. "Introduction. Nomadic Pastoralists and Social Change Processes and Perspectives." Journal of Asian and African Studies 16 (1981c):4-26.

Galaty, J. "Land and Livestock among Kenyan Maasai." Journal of Asian and African Studies 16 (1981d):68-88.

Galbraith, J.K. The Nature of Mass Poverty. Cambridge: Harvard University Press, 1979.

Glover, P., and M. Gwynne. "The Destruction of Maasailand." New Scientist 24 (1961):450-53.

Godelier, M. Rationality and Irrationality in Economics. London: NLB, 1972.

Goldschmidt, W. "A National Livestock Bank: An Institutional Device for Rationalizing the Economy of Tribal Pastoralists." International Development Review 17 (1975):2-6.

Goldschmidt, W. "A General Model for Pastoral Social Systems." Pastoral Production and Society, Proceedings of the International Meeting on Nomadic Pastoralism, Paris, December 1-3, 1976, pp. 15-27. Cambridge: Cambridge University Press, 1979.

Goldschmidt, W. "An Anthropological Approach to Economic Development." Proceedings, Conference on the Future of Pastoral Peoples, Nairobi, Kenya, August 4-8, 1980, ed. J. Galaty, D. Aronson, P. Salzman, and A. Chouinard, pp. 52-58. Ottawa: International Development Research Centre, 1981a.

Goldschmidt, W. "The Failure of Pastoral Economic Development Programs in Africa." Proceedings, Conference on the Future of Pastoral Peoples, Nairobi, Kenya, August 4-8, 1980, ed. J. Galaty, D. Aronson, P. Salzman, and A. Chouinard, pp. 101-18. Ottawa: International Development Research Centre, 1981b.

Grandin, B. "Time Allocation and Labor Inputs on a Maasai Group Ranch--Preliminary Findings from Olkarkar." Prepared for the ILCA Programme Committee, Nairobi, 1982.

Grandin, B., and S. Bekure. "Livestock Offtake and Acquisition: A Preliminary Analysis of Livestock Transactions in Olkarkar and Mbirikani Group Ranches." Prepared for the ILCA Programme Committee, Nairobi, 1982.

Griffen, L., and E. Allonby. "The Economic Effects of Trypanosomiasis in Sheep and Goats at a Range Research Station in Kenya." Tropical Animal Health and Production 11 (1979a):127-32.

Griffen, L., and E. Allonby. "Studies on the Epidemiology of Trypanosomiasis in Sheep and Goats in Kenya." Tropical Animal Health and Production 11 (1979b): 133-42.

Griffiths, J. "The Climate in Kenya Masailand." East African Agricultural and Forestry Journal 28, no. 1 (1962): 1-6.

Grindle, R. "Economic Losses Resulting from Bovine Cysticercosis with Special Reference to Botswana and Kenya." Tropical Animal Health and Production 10 (1978):127-40.

Grootenhuis, J.G., and A.S. Young. "The Importance of Wildlife in the Epidemiology of Theileriosis." Wildlife Disease Research and Economic Development, ed. L. Karstad, B. Nestel, and M. Graham, pp. 33-39. Ottawa: International Research Development Centre, 1981.

Gruchy, A. "Institutional Versus Orthodox Economics and the Interdisciplinary Approach." Economics and Sociology: Towards an Integration, ed. T. Huppes, pp. 51-75. Leiden, Netherlands: Martinus Nijhoff Social Sciences Division, 1976.

Gryseels, G. "Livestock in Farming Systems Research for Smallholder Agriculture: Experiences of ILCA's Highland Programme." Paper presented at the Seminar on Agricultural Research in Rwanda, Kigali, February 5-12, 1983.

Gulliver, P.H. "Nomadic Movements: Causes and Implications." Pastoralism in Tropical Africa, Studies presented and discussed at the XIIIth International African Seminar, Niamey, December 1972, ed. T. Monod, pp. 369-84. London: Oxford University Press, 1975.

Gulliver, P.H. "Letter to the Editor." Africa 48 (1978): 203-04.

Hageboeck, M., G. Cochrane, L. Cooley, and G. Hursh-Cesar. Manager's Guide to Data Collection. Washington, D.C.: U.S.A.I.D. Program Design and Evaluation Methods, November 1979.

Hampson, M. "A Conceptual Approach to Some Land Use Problems in Kajiado District." M.Sc. Thesis, University of Nairobi, 1975.

Hardin, G. "The Tragedy of the Commons." Science 162 (1968):1243-48.

Harrington, G. "Problems of Marketing Beef Cattle in Developed Countries." Beef Cattle Production in Developing Countries, Proceedings of the Conference held in Edinburgh, September 1-6, 1974, ed. A.J. Smith, pp. 406-25. Edinburgh: University of Edinburgh, Centre for Tropical Veterinary Medicine, 1976.

Harwitz, M. "On Improving the Lot of the Poorest: Economic Plans in Kenya." African Studies Review 21, no. 3 (1978):65-73.

Hazlewood, A. "Income Distribution and Poverty--An Unfashionable View." Journal of Modern African Studies 16, no. 1 (March 1978):81-95.

Hazlewood, A. The Economy of Kenya: The Kenyatta Era. Oxford: Oxford University Press, 1979.

Heady, H. Range Management in East Africa. Nairobi, Kenya: The Government Printer, 1960.

Heath, A. "Decision Making and Transactional Theory." Transaction and Meaning, ed. B. Kapferer, pp. 25-40. Philadelphia: Institute for the Study of Human Issues, 1976.

Hedlund, H. "Impact of Group Ranches on a Pastoral Society." University of Nairobi, Institute for Development Studies, Staff Paper No. 100, 1971.

Helland, J. "Sociological Aspects of Pastoral Livestock Production in Africa." Proceedings, First International Rangeland Congress, pp. 79-81, 1978.

Helland, J. "Some Issues in the Study of Pastoralists and the Development of Pastoralism." ILCA (Kenya) Working Document 15, Nairobi, 1980a.

Helland, J. "Social Organization and Water Control among the Borana of Southern Ethiopia." ILCA (Kenya) Working Document 16, Nairobi, 1980b.

Helland, J. "An Outline of Group Ranching in Pastoral Maasai Areas of Kenya." ILCA (Kenya) Working Document 17, Nairobi, 1980c.

Helland, J. "Some Aspects and Implications of the Development of Grazing Blocks in Northeastern Province, Kenya." ILCA (Kenya) Working Document 18, Nairobi, 1980d.

Helland, J. "An Analysis of Afar Pastoralism in the Northeastern Rangelands of Ethiopia." ILCA (Kenya) Working Document 19, Nairobi, 1980e.

Henin, R. "Characteristics and Development Implications of a Fast-Growing Population." Papers on the Kenyan Economy: Performance, Problems and Policies, ed. T. Killick, pp. 193-207. Nairobi, Kenya: Heinemann Educational Books (East Africa) Ltd., 1981.

Heyer, J. "Achievements, Problems and Prospects in the Agricultural Sector." Agricultural Development in Kenya: An Economic Assessment, ed. J. Heyer, J. Maitha, and W. Senga, pp. 1-31. Nairobi, Kenya: Oxford University Press, 1976a.

Heyer, J. "The Marketing System." Agricultural Development in Kenya: An Economic Assessment, ed. J. Heyer, J. Maitha, and W. Senga, pp. 313-63. Nairobi, Kenya: Oxford University Press, 1976b.

Hildebrand, P., and R. Waugh. "Farming Systems Research and Development." Farming Systems Support Project Newsletter 1 (Spring 1983):4-5.

Hinga, S., and J. Heyer. "The Development of Large Farms." Agricultural Development in Kenya: An Economic Assessment, ed. J. Heyer, J. Maitha, and W. Senga, pp. 222-54. Nairobi, Kenya: Oxford University Press, 1976.

Hitchcock, R. "Tradition, Social Justice and Land Reform in Central Botswana." Land Reform in the Making: Tradition, Public Policy and Ideology in Botswana, ed. R. Werbner. Special Number of the Journal of African Law 24 (Spring 1980):1-34.

Hjort, A. "Herds, Trade, and Grain: Pastoralism in a Regional Perspective." Proceedings, Conference on the Future of Pastoral Peoples, Nairobi, Kenya, August 4-8, 1980, ed. J. Galaty, D. Aronson, P. Salzman, and A. Chouinard, pp. 135-42. Ottawa: International Development Research Centre, 1981.

Hogg, R. "Pastoralism and Impoverishment: The Case of the Isiolo Boran of Northern Kenya." Disasters 4 (1980): 299-310.

Hopcraft, D. "A Productivity Comparison between Thomson's Gazelle and Cattle, and Their Relation to the Ecosystem in Kenya." Ph.D. Thesis, Cornell University, 1975.

Hopcraft, P. "Economic Institutions and Pastoral Resources Management: Considerations for a Development Strategy." Proceedings, Conference on the Future of Pastoral Peoples, Nairobi, Kenya, August 4-8, 1980, ed. J. Galaty, D. Aronson, P. Salzman, and A. Chouinard, pp. 224-40. Ottawa: International Development Research Centre, 1981.

Horowitz, M.M. "The Sociology of Pastoralism and African Livestock Projects." Washington, D.C.: U.S.A.I.D. Program Evaluation Discussion Paper No. 6, 1979.

House, W., and T. Killick. "Inequality and Poverty in the Rural Economy, and the Influence of Some Aspects of Policy." Papers on the Kenyan Economy: Performance Problems and Policies, ed. T. Killick, pp. 157-79. Nairobi, Kenya: Heinemann Educational Books (Ease Africa) Ltd., 1981.

Howard, L. "The Role of Wildlife Disease Research in Livestock Production." Wildlife Disease Research and Economic Development, ed. L. Karstad, B. Nestel, and M. Graham, pp. 64-67. Ottawa: International Research Development Centre, 1981.

IBRD. "Kenya Second Livestock Development Project, Credit 477-KE. Review Mission, February 22-March 26, 1976." Nairobi, 1977.

IBRD. Accelerated Development in Sub-Saharan Africa. Washington, D.C.: Author, 1981.

IBRD. World Development Report 1982. New York: Oxford University Press, 1982.

IBRD/IDA. "Study of the Availability of Fattening Beef Breeding and Dairy Breeding Stock in East Africa, 1970-75." IBRD Permanent Mission for Eastern Africa, Nairobi, 1972.

Ikiara, G., and T. Killick. "The Performance of the Economy since Independence." Papers on the Kenyan Economy: Performance, Problems and Policies, ed. T. Killick, pp. 5-19. Nairobi, Kenya: Heinemann Educational Books (East Africa) Ltd., 1981.

ILCA. "Preliminary Results of the Ranch Component Monitoring Programme." ILCA (Kenya), 1978.

ILCA. "Economic Trends: Livestock Production Prospects for Tropical Africa in the Year 2000." Addis Ababa, Ethiopia: ILCA (Headquarters) Bulletin 10, 1980a.

ILCA. "Pastoral Development Projects." Addis Ababa, Ethiopia: ILCA (Headquarters) Bulletin 8, 1980b.

ILCA. "Economic Trends: Small Ruminants." Addis Ababa, Ethiopia: ILCA (Headquarters) Bulletin 7, 1980c.

ILCA. "Introduction to East African Range Livestock Systems Study/Kenya (Draft)." ILCA (Kenya) Working Document 23, Nairobi, 1981.

ILCA/Animal Production Unit (Botswana). "Mathematical Modelling of Livestock Production Systems: Application of the Texas A&M University Beef Cattle Production Model to Botswana." Addis Ababa, Ethiopia: ILCA (Headquarters) Systems Study 1, 1978.

ILO. Yearbook of Labor Statistics, 42nd ed. Geneva: International Labor Organization, 1982.

Ingold, T. Hungers, Pastoralists and Ranchers: Reindeer Economics and Their Transformations. Cambridge: Cambridge University Press, 1980.

Institute for Development Anthropology. "The Workshop on Pastoralism and African Livestock Development." Washington, D.C.: U.S.A.I.D. Program Evaluation Report No. 4, 1980.

International Monetary Fund. International Financial Statistics, Yearbook 1983, vol. 36. Washington, D.C.: Author, 1983.

Jacobs, A.H. "The Pastoral Masai of Kenya." A Report of Anthropological Field Research, submitted to Ministry of Overseas Development, London, 1963.

Jacobs, A.H. "The Traditional Political Organization of the Pastoral Masai." Ph.D. Thesis, Nuffield College, 1965.

Jacobs, A.H. "Maasai Pastoralism in Historical Perspective." Pastoralism in Tropical Africa, Studies presented and discussed at the XIIIth International African Seminar, Niamey, December 1972, ed. T. Monod, pp. 406-23. London: Oxford University Press, 1975.

Jacobs, A.H. "Development in Tanzania Maasailand: The Perspective over 20 Years, 1957-77." A report prepared for U.S.A.I.D. Mission in Tanzania, 1978.

Jacobs, A.H. "Pastoral Development in Tanzanian Maasailand." Rural Africana, no. 7 (Spring 1980a):1-14.

Jacobs, A.H. "Pastoral Maasai and Tropical Rural Development." Agricultural Development in Africa: Issues of Public Policy, ed. R. Bates and M. Lofchie, pp. 275-300. New York: Praeger Publishers, 1980b.

Jaffe, A. "Ranching on the Wild Side." Wildlife International 5 (November-December 1975):5-13.

Jahnke, H.E. "Livestock in Economic Development." ILCA (Kenya) Working Document 11, Nairobi, 1979.

Jahnke, H.E. Livestock Production Systems and Livestock Development in Tropical Africa. Kiel, Germany: Kieler Wissenschaftsverlag Vauk, 1982.

Jahnke, H.E., M. Wales, J-C. Bille, and N. Dyson-Hudson. "ILCA's Monitoring Activities in Kenya: Report to the Government of Kenya." ILCA (Kenya) Working Document 5, Nairobi, 1978.

Jarvis, L. "Cattle as Capital Goods and Ranchers as Portfolio Managers: An Application to the Argentine Cattle Sector." Journal of Political Economy 82 (1974):489-520.

Jarvis, L. "Cattle as a Store of Wealth in Swaziland: Comment." American Journal of Agricultural Economics 62 (1980):606-13.

Jenny, B. "The Administration of Planning and Project Preparation in Kenya." Development Planning in Kenya, ed. T. Pinfold, and G. Norcliffe, pp. 13-32. Downsview, Canada: York University, Geographical Monographs No. 9, 1980.

Johnson, D. "Pastoral Nomadism in the Sahel Zone." Ecological Systems and Ecopolitics: A Reader on Human and Social Implications of Environmental Management in Developing Countries, ed. K. Deutsch, pp. 169-85. Paris: Unesco, 1977.

Johnson, S., and G. Rausser. "Systems Analysis and Simulation: A Survey of Applications in Agricultural and Resource Economics." A Survey of Agricultural Economics Literature, vol. 2, ed. G. Judge, R. Day, S. Johnson, G. Rausser, and L. Martin, pp. 157-301. Minneapolis: University of Minnesota Press, 1977.

Johnston, B.F., and W. Clark. Redesigning Rural Development, A Strategic Perspective. Baltimore: The Johns Hopkins University Press, 1982.

Jones, P.H. "The Marketing of African Livestock." Ministry of Agriculture, Animal Husbandry and Water Resources, Nairobi, 1959.

Josserand, H., and G. Sullivan. Livestock and Meat Marketing in West Africa, vol. 2. Ann Arbor: University of Michigan, Center for Research on Economic Development, 1979.

"Kajiado, District Environmental Assessment Report." National Environment Secretariat, Ministry of Environment and Natural Resources, in cooperation with Clark University and U.S.A.I.D., Nairobi, 1980.

Kaldor, N. "Equilibrium and Growth Theory." Economics and Human Welfare, ed. M. Boskin, pp. 273-91. New York: Academic Press, Inc., 1979.

Kapferer, B. "Introduction: Transactional Models Reconsidered." Transaction and Meaning, ed. B. Kapferer, pp. 1-22. Philadelphia: Institute for the Study of Human Issues, 1976.

Katona, G. "Rational Behavior and Economic Behavior." The Making of Decisions, ed. W. Gore and J. Dyson, pp. 51-63. New York: Free Press of Glencoe, 1964.

Keesing, R. Cultural Anthropology: A Contemporary Perspective. New York: Holt, Rinehart and Winston, Inc., 1976.

Kenworthy, J. "Climate and Development." Studies in East African Geography and Development, ed. S. Ominde, pp. 41-48. Berkeley: University of California Press, 1971.

Kenya, Department of Veterinary Services. "Annual Report, 1975, Rift Valley Province." Nairobi, 1976.

Kenya, Department of Veterinary Services. "Annual Report, 1976, Rift Valley Province." Nairobi, 1977.

Kenya, Department of Veterinary Services. "Annual Report, 1977, Rift Valley Province." Nairobi, 1978.

Kenya, Department of Veterinary Services. "Annual Report, 1978, Rift Valley Province." Nairobi, 1979.

Kenya, Department of Veterinary Services. "Annual Report, 1979, Rift Valley Province." Nairobi, 1980.

Kenya, Ministry of Agriculture. "Progress Report. Brief Marketing Study on Cattle Bought at Garba Tula by Livestock Marketing Division and Slaughtered at K.M.C., Athi River (Draft)." Nairobi, 1970.

Kenya, Ministry of Agriculture. "Report of a Cattle Transport Study Safari to Wajir/Isiolo." Nairobi, 1971.

Kenya, Ministry of Agriculture. "Beef Marketing in Kenya in 1978." Nairobi, 1978.

Kenya, Ministry of Economic Planning and Community Affairs. "Arid and Semi-Arid Lands Development in Kenya. The Framework for Implementation, Programme Planning, and Evaluation." Nairobi, 1979a.

Kenya, Ministry of Economic Planning and Community Affairs. Statistical Abstract 1979. Nairobi: The Government Printer, 1979b.

Kenya, Ministry of Economic Planning and Development. Economic Survey 1980. Nairobi: The Government Printer, 1980a.

Kenya, Ministry of Economic Planning and Development. Statistical Abstract 1980. Nairobi: The Government Printer, 1980b.

Kenya, Republic of. "The Price Control (Meat) Order, 1978." Nairobi: The Government Printer, 1978.

Kenya, Republic of. "The Price Control (Meat) Order, 1979." Nairobi: The Government Printer, 1979.

Kenya, Republic of. National Livestock Development Policy. Nairobi: The Government Printer, 1980a.

Kenya, Republic of. Sessional Paper No. 4 of 1980 on Economic Prospects and Policies. Nairobi: The Government Printer, 1980b.

Kenya, Republic of. "The Strategy of the Fourth Development Plan, 1979-83." Papers on the Kenyan Economy: Performance, Problems and Policies, ed. T. Killick, pp. 90-96. Nairobi, Kenya: Heiemann Educational Books (East Africa) Ltd., 1981a.

Kenya, Republic of. "The Price Control (Meat) (Amendment) Order, 1981." Nairobi: The Government Printer, 1981b.

Kenya, Republic of. "The Price Control (Meat) Order, 1983." Nairobi: The Government Printer, 1983.

Killick, T. "Retail Price Controls: Some Elementary Theory and Some African Experiences." Presented at a Workshop held at the Institute for Development Studies of the University of Nairobi, March 26-29, 1979. Printed in Price and Marketing Controls in Kenya, ed. J.T. Mukui, pp. 12-20. Institute for Development Studies, Occasional Paper No. 32, 1979.

King, J.M., and B.R. Heath. "Game Domestication for Animal Production in Africa." World Animal Review 16 (1975): 23-30.

King, J.M., A.R. Sayers, C. Peacock, and E. Kontrohr. "Maasai Herd and Flock Structure in Relation to Household Livestock Wealth and Group Ranch Development (Draft)." ILCA (Kenya) Working Document 27, Nairobi, 1982.

King, R. Land Reform. Boulder, Colorado: Westview Press, 1977.

Kivunja, C. The Economics of Cattle and Beef Marketing in Kenya. Frankfurt am Main, Germany: DLG-Verlag, 1978.

Konandreas, P.A., and F.M. Anderson. "Cattle Herd Dynamics: An Integer and Stochastic Model for Evaluating Production Alternatives (Draft)." Addis Ababa, Ethiopia: ILCA (Headquarters), 1980.

Konczacki, Z. The Economics of Pastoralism. London: Frank Cass, 1978.

Krostitz, W. "The New International Market for Game Meat." Unasylva 31, no. 123 (1979):32-36.

Laksesvela, B., and A. Said. "Tropical Versus Pemperate Grasses." World Review of Animal Production 14, no. 3 (1978):49-57.

Laughlin, T. "Environment, Political Culture, and National Orientations: A Comparison of Settled and Pastoral Masai." Values, Identities, and National Integration, ed. J. Paden, pp. 91-103. Evanston: Northwestern University Press, 1980.

Lawry, S. "Land Tenure, Land Policy, and Smallholder Livestock Development in Botswana." University of Wisconsin-Madison, Land Tenure Center, Research Paper No. 78, March 1983.

Lele, U. The Design of Rural Development: Lessons from Africa. Baltimore: The Johns Hopkins University Press, 1975.

Levine, J., and W. Hohenboken. "Modelling of Beef Production Systems." World Animal Review 43 (1982):33-40.

Lewis, I. "The Dynamics of Nomadism: Prospects for Sedentarization and Social Change." Pastoralism in Tropical Africa, Studies presented and discussed at the XIIIth International African Seminar, Niamey, December 1972, ed. T. Monod, pp. 426-40. London: Oxford University Press, 1975.

Lindblom, C. "The Science of 'Muddling Through.'" The Making of Decisions, ed. W. Gore and J. Dyson, pp. 155-69. New York: Free Press of Glencoe, 1964.

Little, P.D. "Food Production, Marketing and Consumption in the Semi-Arid Area of Baringo District." Nairobi, Kenya: A Report Prepared for the Nutrition Planning Unit, Ministry of Economic Planning and Development (Kenya), and FAO, 1980.

Little, P.D. "Risk Aversion, Economic Diversification and Goat Production: Some Comments on the Role of Goats in African Pastoral Production Systems." Proceedings, Third International Conference on Goat Production and Disease, pp. 428-30. Scottsdale, Arizona: Dairy Goat Journal, 1982a.

Little, P.D. "The Workshop on Development and African Pastoral Livestock Production." Binghamton, New York: Institute for Development Anthropology, U.S.A.I.D. Number AFR-0085-C-00-1033, 1982b.

Little, P.D. "Critical Socio-Economic Variables in African Pastoral Livestock Development: Toward a Comparative Framework." Livestock Development in Subsaharan Africa: Constraints, Prospects, Policy, ed. J.R. Simpson and P. Evangelou, pp. 201-14. Boulder, Colorado: Westview Press, 1984.

Livingstone, I. "Improvements in Kenya's Livestock Economy: Lessons from the S.R.D.P." University of Nairobi, Institute for Development Studies, Working Paper No. 226, 1975.

Livingstone, I. "Cowboys in Africa, the Socio-Economics of Ranching." University of Nairobi, Institute for Development Studies, Occasional Paper No. 17, 1976.

Livingstone, I. "Economic Irrationality among Pastoral Peoples in East Africa: Myth or Reality?" University of Nairobi, Institute for Development Studies, Discussion Paper No. 245, 1977.

Livingstone, I. "The Socio-Economics of Ranching in Kenya." Research in Economic Anthropology 2 (1979):361-90.

Livingstone, I. "Experimentation in Rural Development: The Special Rural Development Programme." Papers on the Kenyan Economy: Performance, Problems and Policies, ed. T. Killick, pp. 320-28. Nairobi, Kenya: Heinemann Educational Books (East Africa) Ltd., 1981.

Lofchie, M., and S. Commins. "Food Deficits and Agricultural Policies in Troipcal Africa." Journal of Modern African Studies 20 (1982):1-25.

Low, A., R. Kemp, and M. Doran. "Cattle as a Store of Wealth in Swaziland: Reply." American Journal of Agricultural Economics 62 (1980):614-17.

Lynam, J. "An Analysis of Population Growth, Technical Change, and Risk in Peasant, Semi-Arid Farming Systems: A Case Study of Machakos District, Kenya." Ph.D. Thesis, Stanford University, 1978.

MacCrimmon, K. "An Overview of Multiple Objective Decision Making." Multiple Criteria Decision Making, ed. J. Cochrane and M. Zeleny, pp. 18-44. Columbia: University of South Carolina Press, 1973.

Malechek, J.C. "Grazing Management of Goats in Extensive Rangeland Production Systems." Proceedings, Third International Conference on Goat Production and Disease, pp. 404-08. Scottsdale, Arizona: Dairy Goat Journal, 1982.

Maloiy, G., and H. Heady. "Grazing Conditions in Kenya Masailand." Journal of Range Management 18 (1965): 269-72.

Marimi, A.M. "Dryland Farming in Kenya: Problems and Prospects." Paper prepared for the Workshop on the Development of Kenya's Semi-Arid Areas, Institute for Development Studies, Nairobi, July 23-26, 1979.

Matthes, M.C. "Livestock Marketing in Kenya." Ministry of Agriculture, Marketing Development Project, Phase II (Ken 75-005), Field Draft Document, Nairobi, 1979.

Matthes, M.C. "Yields of Special Cuts by Grade." Ministry of Agriculture, Nairobi, 1980.

Mbithi, P. "Human Factors in Agricultural Management in East Africa." Food Policy 2 (February 1977):27-33.

McArthur, I.D., and C. Smith. "Price and Marketing Policies on Meat and Eggs." Presented at a Workshop held at the Institute for Development Studies of the University of Nairobi, March 26-29, 1979. Printed in Price and Marketing Congrols in Kenya, ed. J.T. Mukui, pp. 205-23. Institute for Development Studies, Occasional Paper No. 32, 1979.

McCauley, E.H. "Lessons from a Survey of Successful Experiences in Assisting the Smallholder Livestock Producer." World Animal Review 17 (1976):28-33.

McCauley, E.H. "Animal Diseases in Developing Countries: Technical and Economic Aspects of Their Impact and Control." Washington, D.C.: World Bank, AGR Technical Note No. 7, 1983.

McChesney, A. "The Promotion of Economic and Political Rights: Two African Approaches." Journal of African Law 24 (1980):163-205.

McCown, R., G. Haaland, and C. DeHaan. "The Interaction Between Cultivation and Livestock Production in Semi-Arid Africa." Agriculture in Semi-Arid Environments, ed. A. Hall, H. Cannell, and H. Lawton. Berlin: Springer-Verlag, Ecological Studies, 34, 1979.

McDowell, R. "Livestock Nutrition in Subsaharan Africa: An Overview." Livestock Development in Subsaharan Africa: Constraints, Prospects, Policy, ed. J.R. Simpson and P. Evangelou, pp. 43-59. Boulder, Colorado: Westview Press, 1984.

Meadows, S.J., and J.M. White. "Structure of the Herd and Determinants of Offtake Rates in Kajiado District in Kenya 1962-1977." London: Overseas Development Institute, Pastoral Network Paper 7d, March 1979.

Meadows, S.J., and J.M. White. "Cattle Herd Structures in Kenya's Pastoral Rangelands." London: Overseas Development Institute, Pastoral Network Paper 11e, January 1981.

Meier, G., ed. Pricing Policy for Development Management. Published for the Economic Development Institute of the World Bank. Baltimore: The Johns Hopkins University Press, 1983.

Memon, P. "Kenya." African Perspectives, ed. H. de Blij and E. Martin, pp. 59-89. New York: Methuen, 1981.

Meyn, K. "Livestock Constraints." Paper presented at the Workshop sponsored by ILCA on The Design and Implementation of Pastoral Development Projects for Tropical Africa, Addis Ababa, February 25-29, 1980. Addis Ababa, Ethiopia: ILCA (Headquarters) Working Document 4, pp. 73-85, 1980.

Meyn, K., and D. Mbogo. "The Use of Artificial Insemination in Genetic Improvement Programmes for Beef Cattle in Developing Countires." Beef Cattle Production in Developing Countries, Proceedings of the Conference held in Edinburgh, September 1-6, 1974, ed. A.J. Smith, pp. 347-57. Edinburgh: University of Edinburgh, Centre for Tropical Veterinary Medicine, 1976.

Meyn, K., and J.V. Wilkins. "Breeding for Milk in Kenya, with Particular Reference to the Sahiwal Stud." World Animal Review 11 (1974):24-30.

Migot-Adholla, S., and P.D. Little. "Evolution of Policy toward the Development of Pastoral Areas in Kenya." Proceedings, Conference on the Future of Pastoral Peoples, Nairobi, Kenya, August 4-8, 1980, ed. J. Galaty, D. Aronson, P. Salzman, and A. Chouinard, pp. 144-53. Ottawa: International Development Research Centre, 1981.

Mittendorf, H.J. "Factors Affecting the Location of Slaughterhouses in Developing Countries." World Animal Review 25 (1978):13-17.

Molnar, J., and H. Clonts. "Technology as a Source of Economic and Social Advancement in Developing Countries." Transferring Food Production Technology to Developing Nations, ed. J. Molnar and H. Clonts, pp. 1-15. Boulder, Colorado: Westview Press, 1983.

Morris, W. "Farming Systems Research in West African Agricultural Development." Prepared under U.S.A.I.D. Contracts AFR-C-1257, 1258, and 1472 for the Farming Systems Workshop, Dakar, Senegal, January 12-15, 1981.

Mosher, A.T. Creating a Progressive Rural Structure to Serve a Modern Agriculture. New York: Agricultural Development Council, 1969.

Murray, M., W.I. Morrison, P.K. Murray, D.J. Clifford, and J.C.M. Trail. "Trypanotolerance--A Review." World Animal Review 31 (1979):2-12.

Murray, M., J.C.M. Trail, Y. Wissocq, and A.D. Wilson. "Trypanotolerance and Potential Economic Benefits." Presented at the Seminar on Epidemiology and Economics of Trypanosomiasis Control in Selected Areas of Kenya, June 18-19, 1981, Nairobi. Printed in Livestock Productivity and Trypanotolerance, ILCA (Kenya) Working Document 22, 1981.

Mushi, E., F. Rurangirwa, and L. Karstad. "Epidemiology and Control of Bovine Malignant Catarrhal Fever." Wildlife Disease Research and Economic Development, ed. L. Karstad, B. Nestel, and M. Graham, pp. 21-23. Ottawa: International Research Development Centre, 1981.

Naga, M., and K. El-Shazly. "Use of By-Products in Animal-Feeding Systems in the Delta of Egypt." By-Product Utilization for Animal Production, ed. B. Kiflewahid, G. Potts, and R. Drysdale, pp. 9-15. Ottawa: International Development Research Centre, 1983.

Newberry, D. "A Feasibility Study of Intensive Cattle Breeding." Project Appraisal in Practice, ed. M. Scott, J. MacArthur, and D. Newberry, pp. 413-534. London: Heinemann Educational Books, 1976.

Ngulo, W. "Implementation of Livestock Development Programmes. Livestock Development Programmes in Kenya." Paper presented at the Fourth World Conference on Animal Production, Buenos Aires, 1978. Published in Proceedings 1 (1980):202-07.

Njiru, G.K. "The Rendille Economy--Some Preliminary Findings on Trading Activities." University of Nairobi, Institute for Development Studies, Working Paper No. 389, 1982.

Njoka, T. "Ecological and Socio-Cultural Trends of Kaputiei Group Ranches in Kenya." Ph.D. Thesis, University of California, Berkeley, 1979.

Njoka, T. "The Trend of Small Ruminants Population in Masai Group Ranches (1967-1977)--With Special Reference to Kaputiei Section, Kajiado District." Proceedings of the Second Small Ruminant CRSP Kenya Workshop, Naivasha, Kenya, February 22, 1983, pp. 229-44. Prepared by the Management Entity Office, Davis, California, 1983.

Norman, D. "The Farming Systems Approach to Research." Presented at the Farming Systems Research Symposium on Farming Systems in the Field, Kansas State University, Manhattan, 1982.

Norris, J.J. "Desertification and the Goat." Proceedings, Third International Conference on Goat Production and Disease, pp. 409-10. Scottsdale, Arizona: Dairy Goat Journal, 1982.

Odingo, R. "Settlement and Rural Development in Kenya." Studies in East African Geography and Development, ed. S. Ominde, pp. 162-76. Berkeley: University of California Press, 1971.

O'Donovan, P. "Fattening Crossbred and Zebu Cattle on Local Feeds and By-Products in Ethiopia." World Animal Review 30 (1979):23-29.

Oliver, R., and A. Atmore. Africa since 1800, 3rd ed. Cambridge: Cambridge University Press, 1981.

Ominde, S. Land and Population Movements in Kenya. London: Heinemann Educational Books, Ltd., 1968.

Ominde, S. "The Semi-Arid and Arid Lands of Kenya." Studies in East African Geography and Development, ed. S. Ominde, pp. 146-61. Berkeley: University of California Press, 1971.

Ominde, S. "The Integration of Environmental and Development Planning for Ecological Crisis Areas in Africa." Ecological Systems and Ecopolitics: A Reader on Human and Social Implications of Environmental Management in Developing Countries, ed. K. Deutsch, pp. 115-30. Paris: Unesco, 1977.

Ominde, S. "Regional Disparities and the Employment Problem in Kenya." The Spatial Structure of Development: A Study of Kenya, ed. R. Obudho and D. Taylor, pp. 46-73. Boulder, Colorado: Westview Press, 1979.

Oxby, C. "Group Ranches in Africa." World Animal Review 42 (1982):11-18.

Palmquist, R., and E. Pasour, Jr. "Common Property Externalities: Isolation, Assurance, and Resource Depletion in a Traditional Grazing Context: Comment." American Journal of Agricultural Economics 64 (1982):783-84.

Parsons, T., and N. Smelser. Economy and Society. New York: The Free Press, 1956.

Payne, W. "Systems of Beef Production in Developing Countries." Beef Cattle Production in Developing Countries, Proceedings of the Conference held in Edinburgh, September 1-6, 1974, ed. A.J. Smith, pp. 118-31. Edinburgh: University of Edinburgh, Centre for Tropical Veterinary Medicine, 1976.

Peacock, C. "A Preliminary Report of Goat and Sheep Productivity on Olkarkar, Merueshi and Mbirikani Group Ranches." Prepared for the ILCA Programme Committee, Nairobi, 1982.

Peacock, C., P.N. de Leeuw, and J.M. King. "Herd Movement in the Mbirikani Area." Prepared for the ILCA Programme Committee, Nairobi, 1982.

Peberdy, J. "Rangeland." East Africa: Its Peoples and Resources, ed. W. Morgan, pp. 153-76. London: Oxford University Press, 1969.

Porter, P. "Environmental Potential and Economic Opportunities: A Background for Cultural Adaptation." American Anthropologist 67 (1965):409-20.

Potts, G. "Application of Research Results on By-Product Utilization: Economic Aspects to Be Considered." By-Product Utilization for Animal Production, ed. B. Kiflewahid, G. Potts, and R. Drysdale, pp. 116-27. Ottawa: International Development Research Centre, 1983.

Pratt, D. "The Ecological Management of Arid and Semi-Arid Rangeland in Africa and the Near East: The Concept of 'Discrete Development Areas' as Applied to Range Development." Prepared for the FAO, mimeographed, n.d.

Pratt, D. "Design by Objective." Paper presented at the Workshop sponsored by ILCA on The Design and Implementation of Pastoral Development Projects for Tropical Africa, Addis Ababa, February 25-29, 1980. Addis Ababa, Ethiopia: ILCA (Headquarters) Working Document 4, pp. 109-25, 1980.

Pratt, D., and M. Gwynne, ed. Rangeland Management and Ecology in East Africa. London: Hodder and Stoughton, 1977.

Preston, T.R. "Prospects for the Intensification of Cattle Production in Developing Countries." Beef Cattle Production in Developing Countries, Proceedings of the Conference held in Edinburgh, September 1-6, 1974, ed. A.J. Smith, pp. 242-57. Edinburgh: University of Edinburgh, Centre for Tropical Veterinary Medicine, 1976.

Preston, T.R. "A Strategy for Cattle Production in the Tropics." World Animal Review 21 (1977):11-18.

Purcell, W. Agricultural Marketing: Systems, Coordination, Cash and Futures Prices. Reston, Virginia: Reston Publishing Company, 1979.

Quam, M. "Cattle Marketing and Pastoral Conservatism: Karamoja District, Uganda, 1948-1970." African Studies Review 21, no. 1 (April 1978):49-71.

Rae, A. Crop Management Economics. London: Granada Publishing Limited, 1977.

Ranjhan, S. "Use of Agro-Industrial By-Products in Feeding Ruminants in India." World Animal Review 28 (1978): 31-37.

Rennie, T., D. Light, A. Rutherford, M. Miller, I. Fisher, D. Pratchett, B. Capper, N. Buck, and J. Trail. "Beef Cattle Productivity under Traditional and Improved Management in Botswana." Tropical Animal Health and Production 9 (1977):1-6.

Reul, R. "Productive Potential of Wild Animals in the Tropics." World Animal Review 32 (1979):18-24.

Richards, G. "Planning for the Future Development of the Tourist Sector in Kenya." Development Planning in Kenya, ed. T. Pinfold and G. Norcliffe, pp. 141-56. Downsview, Canada: York University, Geographical Monographs No. 9, 1980.

Rimmer, D. "The Economic Imprint of Colonialism and Domestic Food Supplies in British Tropical Africa." Imperialism, Colonialism, and Hunger: East and Central Africa, ed. R. Rotberg, pp. 141-65. Lexington, Massachusetts: D.C. Heath and Company, 1983.

Riney, T. "Wildlife vs. Nomadic Stocks." Unasylva 31, no. 124 (1979):15-20.

Roe, E., and L. Fortmann. "Season and Strategy: The Changing Organization of the Rural Water Sector in Botswana." Ithaca: Cornell University, Rural Development Committee, Special Series on Resource Management, RM No. 1, 1982.

Rogers, L. "The Cultural and Economic Aspects of Range Livestock Production." Proceedings: Integrated Range/Livestock Workshop, ed. J. Henson, J. Noel, and B. Patrick, pp. 20-31. Tucson: Consortium for International Development, 1983.

Roy, A. "Safety First and the Holding of Assets." Econometrica 20 (1952):431-49.

Runge, C. "Common Property Externalities: Isolation, Assurance, and Resource Depletion in a Traditional Grazing Context." American Journal of Agricultural Economics 63 (1981):595-606.

Runge, C. "Common Property Externalities: Isolation, Assurance, and Resource Depletion in a Traditional Grazing Context: Reply." American Journal of Agricultural Economics 64 (1982):785-88.

Ruthenberg, H. African Agricultural Production Development Policy in Kenya 1952-1965. Berlin: Springer-Verlag, 1966.

Ruthenberg, H. "Economic Objectives in Pastoral Projects." Paper presented at the Workshop sponsored by ILCA on The Design and Implementation of Pastoral Development Projects for Tropical Africa, Addis Ababa, February 25-29, 1980. Addis Ababa, Ethiopia: ILCA (Headquarters) Working Document 4, pp. 15-35, 1980a.

Ruthenberg, H. Farming Systems in the Tropics, 3rd ed. Oxford: Clarendon Press, 1980b.

Said, A.N., F. Sundstol, S. Tubei, N. Musimba, and F. Ndegwa. "Use of By-Products for Ruminant Feeding in Kenya." By-Product Utilization for Animal Production, ed. B. Kiflewahid, G. Potts, and R. Drysdale, pp. 60-70. Ottawa: International Development Research Centre, 1983.

Saitoti, T. Maasai. London: Elm Tree Books, 1980.

Salzman, P. "Introduction: Processes of Sedentarization as Adaptation and Response." When Nomads Settle, ed. P. Salzman, pp. 1-19. New York: Praeger Publishers, 1980.

Salzman, P. "Political Factors in the Future of Pastoral Peoples." Proceedings, Conference on the Future of Pastoral Peoples, Nairobi, Kenya, August 4-8, 1980, ed. J. Galaty, D. Aronson, P. Salzman, and A. Chouinard, pp. 130-33. Ottawa: International Development Research Centre, 1981a.

Salzman, P. "Afterword: On Some General Theoretical Issues." Journal of Asian and African Studies 16 (1981b): 158-66.

Samuelson, P. Economics, 9th ed. New York: McGraw-Hill, 1973.

Sandford, S. "Organizing Government's Role in the Pastoral Sector." Proceedings, Conference on the Future of Pastoral Peoples, Nairobi, Kenya, August 4-8, 1980, ed. J. Galaty, D. Aronson, P. Salzman, and A. Chouinard, pp. 270-79. Ottawa: International Development Research Centre, 1981.

Sandford, S. "Institutional and Economic Issues in the Development of Goat and Goat Product Markets." Proceedings, Third International Conference on Goat Production and Disease, pp. 31-35. Scottsdale, Arizona: Dairy Goat Journal, 1982.

Sandford, S. Management of Pastoral Development in the Third World. Chichester, England: John Wiley & Sons, 1983.

Schaefer-Kehnert, W. "Price Policy." Paper S 12/68-18, Presented at the Seminar on Animal Husbandry and Marketing, German Foundation for Developing Countries, Nairobi, Kenya, September 30-October 16, 1968.

Schaefer-Kehnert, W. "Economic Aspects of Intensive Beef Cattle Feeding in Kenya." UNDP/FAO Beef Industry Development Project, Nairobi, 1971.

Schaefer-Kehnert, W. "Economic Aspects of Intensive Beef Production in a Developing Country." Quarterly Journal of International Agriculture 17 (October-December 1978): 342-52.

Schaefer-Kehnert, W. "Appraisal and Finance of Intensive Animal Production Schemes." Paper presented at the Conference on Intensive Animal Production in Developing Countries, Harrogate, England, 1979.

Schaefer-Kehnert, W., and L. Brown. "Economic and Social Aspects of Animal Production in Relation to Conservation and Recreation." Paper presented at the Third World Conference on Animal Production, Melbourne, 1973. Published in Proceedings, Theme 1, Paper 5 (1973): 47-52.

Scherer, F. Industrial Market Structure and Economic Performance. Chicago: Rand McNally College Publishing Co., 1970.

Schneider, H. Livestock and Equality in East Africa: The Economic Basis for Social Structure. Bloomington: Indiana University Press, 1979.

Schneider, H. "Livestock as Food and Money." Proceedings, Conference on the Future of Pastoral Peoples, Nairobi, Kenya, August 4-8, 1980, ed. J. Galaty, D. Aronson, P. Salzman, and A. Chouinard, pp. 210-19. Ottawa: International Development Research Centre, 1981a.

Schneider, H. "The Pastoralist Development Problem." Journal of Asian and African Studies 16 (1981b):27-32.

Schultz, T. Transforming Traditional Agriculture. New Haven: Yale University Press, 1964.

Scoville, O., and M. Sarhan. "Objectives and Constraints of Ruminant Livestock Production." World Review of Animal Production, 14, no. 1 (1978):43-48.

Semenye, P.P. "A Study of Masai Herds on Elangata Wuas Group Ranch in Kajiado District, Kenya." ILCA (Kenya) Working Document 14, Nairobi, 1980.

Semenye, P.P. "A Preliminary Report of Cattle Productivity in Olkarkar, Merueshi and Mbirikani Group Ranches." Prepared for the ILCA Programme Committee, Nairobi, 1982.

Semenye, P.P., and F.N. Chabari. "Ranch Performance under the Kenya Livestock Development Project: A Herd Performance and Financial Assessment of 10 Ranches (Revised)." ILCA (Kenya) Working Document 21, Nairobi, 1980.

Senga, W. "Kenya's Agricultural Sector." Agricultural Development in Kenya: An Economic Assessment, ed. J. Heyer, J. Maitha, and W. Senga, pp. 69-110. Nairobi, Kenya: Oxford University Press, 1976.

Sfeir-Younis, A., and D. Bromley. Decision Making in Developing Countries. New York: Praeger Publishers, 1977.

Shah, S., and Z. Muller. "Feeding Animal Wastes to Ruminants." By-Product Utilization for Animal Production, ed. B. Kiflewahid, G. Potts, and R. Drysdale, pp. 49-57. Ottawa: International Development Research Centre, 1983.

Shaner, W., P. Philipp, and W. Schmehl. Farming Systems Research and Development: Guidelines for Developing Countries. Boulder, Colorado: Westview Press, 1982.

Sharpley, J. "Resource Transfers between the Agricultural and Nonagricultural Sectors: 1964-77." Papers on the Kenyan Economy: Performance, Problems and Policies, ed. T. Killick, pp. 311-19. Nairobi, Kenya: Heinemann Educational Books (East Africa) Ltd., 1981.

Siebert, H. Economics of the Environment. Lexington, Massachusetts: D.C. Heath and Company, 1981.

Simon, H. "A Behavioral Model of Rational Choice." Quarterly Journal of Economics 69 (1955):99-118.

Simpson, J.R. "An Economic Evaluation of Selected Range Improvement Practices on the Papago Indian Reservation." M.S. Thesis, University of Arizona, 1968.

Simpson, J.R. "Uses of Cultural Anthropology in Economic Analysis: A Papago Indian Case." Human Organization 29 (Fall 1970):162-68.

Simpson, J.R. "Identification of Goals and Strategies in Designing Technological Change for Developing Countries." Transferring Food Production Technology to Developing Nations, ed. J. Molnar and H. Clonts, pp. 29-41. Boulder, Colorado: Westview Press, 1983.

Simpson, J.R. "Problems and Constraints, Goals and Policy: Conflict Resolution in Development of Subsaharan Africa's Livestock Industry." Livestock Development in Subsaharan Africa: Constraints, Prospects, Policy, ed. J.R. Simpson and P. Evangelou, pp. 5-20. Boulder, Colorado: Westview Press, 1984.

Simpson, J.R., and D. Farris. The World's Beef Business. Ames, Iowa: The Iowa State University Press, 1982.

Simpson, J.R., and J. Mirowsky. "World Trade in Canned Beef, 1962-1976." Gainesville: University of Florida, Center for Tropical Agriculture, Report 2, November 1979.

Simpson, M.C. "Problems of Marketing Beef Cattle in Developing Countries." Beef Cattle Production in Developing Countries, Proceedings of the Conference held in Edinburgh, September 1-6, 1974, ed. A.J. Smith, pp. 426-29. Edinburgh: University of Edinburgh, Centre for Tropical Veterinary Medicine, 1976.

Skovlin, J. "Ranching in East Africa: A Case Study." Journal of Range Management 24, no. 4 (July 1971): 263-70.

Slovic, P., H. Kunreuther, and G. White. "Decision Processes, Rationality, and Adjustment to Natural Hazards." Natural Hazards, ed. G. White, pp. 187-205. New York: Oxford University Press, 1974.

Smelser, N. "On the Relevance of Economic Sociology for Economics." Economics and Sociology: Towards an Integration, ed. T. Huppes, pp. 1-26. Leiden, Netherlands: Martinus Nijhoff Social Sciences Division, 1976.

Smith, L. "An Overview of Agricultural Development Policy." Agricultural Development in Kenya: An Economic Assessment, ed. J. Heyer, J. Maitha, and W. Senga, pp. 111-51. Nairobi, Kenya: Oxford University Press, 1976.

Spedding, C. "The Study of Agricultural Systems." Study of Agricultural Systems, ed. G. Dalton, pp. 3-19. London: Applied Science Publishers, Ltd., 1975.

Spencer, I. "Settler Dominance, Agricultural Production and the Second World War in Kenya." Journal of African History 21 (1980):497-514.

Spencer, I. "Pastoralism and Colonial Policy in Kenya, 1895-1929." Imperialism, Colonialism, and Hunger: East and Central Africa, ed. R. Rotberg, pp. 113-40. Lexington, Massachusetts: D.C. Heath and Company, 1983.

Spink, B.D., B. Persson, J. Gardell, and P. Horelli. "An Organizational and Management Review of the Kenya Meat Commission." Rome: FAO Draft Consultant Report, 1980.

Squire, H. "Phased Development of Masai Feedlot Proposal, Kenya Breweries Farm, Mau Narok." Beef Research Station, Lanet, Kenya, 1975.

Squire, H. "Experiences with the Development of an Intensive Beef Feeding System in Kenya." Beef Cattle Production in Developing Countries, Proceedings of the Conference held in Edinburgh, September 1-6, 1974, ed. A.J. Smith, pp. 150-53. Edinburgh: University of Edinburgh, Centre for Tropical Veterinary Medicine, 1976.

Stanfield, J. Economic Thought and Social Change. Carbondale, Illinois: Southern Illinois University Press, 1979.

Stevenson, P., A. Jones, and L. Khalil. "The Public Health Significance of Cysticercosis in African Game Animals." Wildlife Disease Research and Economic Development, ed. L. Karstad, B. Nestel, and M. Graham, pp. 57-61. Ottawa: International Research Development Centre, 1981.

Stewart, F. "Kenya: Strategies for Development." Papers on the Kenyan Economy: Performance, Problems and Policies, ed. T. Killick, pp. 75-89. Nairobi, Kenya: Heinemann Educational Books (East Africa) Ltd., 1981.

Stiles, D. "Relevance of the Past in Projections about Pastoral Peoples." Proceedings, Conference on the Future of Pastoral Peoples, Nairobi, Kenya, August 4-8, 1980, ed. J. Galaty, D. Aronson, P. Salzman, and A. Chouinard, pp. 370-78. Ottawa: International Development Research Centre, 1981.

Street, D., and G. Sullivan. "Marketing Decisions for Small-Scale Producers." Transferring Food Production Technology to Developing Nations, ed. J. Molnar and H. Clonts, pp. 106-23. Boulder, Colorado: Westview Press, 1983.

Stryker, J.D. "Land Use Development in the Pastoral Zone of West Africa." Livestock Development in Subsaharan Africa: Constraints, Prospects, Policy, ed. J.R. Simpson and P. Evangelou, pp. 175-85. Boulder, Colorado: Westview Press, 1984.

Sullivan, G. "Economics of Improved Management for Transforming the Forage/Livestock System in Tanzania--A Simulation Model." Ph.D. Thesis, Texas A&M University, 1979.

Sullivan, G. "Impact of Government Policies on the Performance of the Livestock-Meat Subsector." Livestock Development in Subsaharan Africa: Constraints, Prospects, Policy, ed. J.R. Simpson and P. Evangelou, pp. 143-59. Boulder, Colorado: Westview Press, 1984.

Sullivan, G., D. Farris, and J.R. Simpson. "Livestock Management Systems in East Africa: An Alternative to Uncontrolled Communal Grazing." Paper presented at the American Agricultural Economics Association Meetings held in Logan, Utah, August 2-5, 1982.

Sullivan, G., D. Farris, M. Yetley, and W. Njukia. "A Socio-Economic Analysis of Technology Adoption in an African Livestock Industry." Texas Agricultural Experiment Station, Technical Article 13513, Texas A&M University, 1978.

Sullivan, G., K. Stokes, D. Farris, T. Nelson, and T. Cartwright. "Transforming a Traditional Forage/Livestock System to Improve Human Nutrition in Tropical Africa." Journal of Range Management 33 (May 1980):174-78.

Surujbally, R. "Game Farming Is a Reality." Unasylva 29, no. 116 (1977): 13-16.

Swift, J. "Pastoral Nomadism as a Form of Land-Use: The Twareg of the Adrar n Iforas." Pastoralism in Tropical Africa, Studies presented and discussed at the XIIIth International African Seminar, Niamey, December 1972, ed. T. Monod, pp. 443-53. London: Oxford University Press, 1975.

Swift, J. "The Development of Livestock Trading in Nomad Pastoral Economy: The Somali Case." Pastoral Production and Society, Proceedings of the International Meeting on Nomadic Pastoralism, Paris, December 1-3, 1976, pp. 447-65. Cambridge: Cambridge University Press, 1979.

Swift, J. "Rapid Appraisal and Cost-Effective Participatory Research in Dry Pastoral Areas of West Africa." Agricultural Administration 8 (1981):485-92.

Swift, J. "The Future of African Hunter-Gatherer and Pastoral Peoples." Development and Change 13 (April 1982):159-81.

Talbot, L. "Ecological Consequences of Rangeland Development in Masailand, East Africa." The Careless Technology: Ecology and International Development, The Record of the Conference on the Ecological Aspects of International Development, Washington University, Warrenton, Virginia, December 8-11, 1968, ed. M. Farvar and J. Milton, pp. 694-711. Garden City, New York: The Natural History Press, 1972.

Thompson, K., R. Todorovic, G. Mateus, and L. Adams. "Methods to Improve the Health of Cattle in the Tropics: Conclusions and Economic Appraisal." Tropical Animal Health and Production 10 (1978):141-44.

Thresher, P. "The Economics of Domesticated Oryx Compared with That of Cattle." World Animal Review 36 (1980): 37-43.

Thresher, P. "The Present Value of an Amboseli Lion." World Animal Review 40 (1981):30-33.

Tignor, R. The Colonial Transformation of Kenya. Princeton: Princeton University Press, 1976.

Trail, J.C.M. "Merits and Demerits of Importing Exotic Cattle Compared with Improvement of Local Breeds: Cattle in Africa South of the Sahara." Presented at the Conference on Intensive Animal Production in Developing Countries, November, 1979, Harrowgate, England. Printed in Livestock Productivity and Trypanotolerance, ILCA (Kenya) Working Document 22, 1981.

Trail, J.C.M., and K.E. Gregory. "Sahiwal Cattle in Africa: How They Can Help Improve Milk and Beef Production." Span 24 (1981a):28-30.

Trail, J.C.M., and K.E. Gregory. "Sahiwal Cattle: An Evaluation of Their Potential Contribution to Milk and Beef Production in Africa." Addis Ababa, Ethiopia: ILCA Monograph 3, 1981b.

UN, Department of International Economic and Social Affairs. Demographic Yearbook, 1981. New York: United Nations Publishing Service, 1981.

UNDP/FAO. "Pastoral Herd Offtake. Its Significance, a Procedure to Establish It, and to Make Future Projections." Prepared for the Ministry of Agriculture (Kenya), and based on the work of H.R. Keymeulen, n.d.

UNDP/FAO. "Policy Issues for Livestock Marketing." Ministry of Agriculture, Marketing Development Project, Phase II (Ken 78-006), Discussion Paper, Nairobi, 1979.

UNDP/FAO. "Meat Wholesale Marketing in Nairobi." Ministry of Agriculture, Marketing Development Project, Phase II (Ken 78-006), Nairobi, 1980.

Unesco/UNEP/FAO. Tropical Grazing Land Ecosystems. Paris: Unesco, 1979.

von Kaufmann, R. "The Development of the Range Land Areas." Agricultural Development in Kenya: An Economic Assessment, ed. J. Heyer, J. Maitha, and W. Senga, pp. 255-87. Nairobi, Kenya: Oxford University Press, 1976.

Wales, M. "Proposed Programmes for Monitoring the Kenya Livestock Development Project." ILCA (Kenya) Working Document 1, Nairobi, 1978.

Wales, M., and F.N. Chabari. "District Ranch Development Briefs." ILCA (Kenya) Working Document 13, Nairobi, 1979.

Waller, R. "Comment on Network Paper 7d/1979 (Meadows and White on: Herd Structure and Offtake in Kajiado, Kenya)." London: Overseas Development Institute, Pastoral Network Paper 9e, December 1979.

Walters, H. "Agriculture and Development." Finance and Development 19 (September 1982):6-11.

Watson, R.M. "Aerial Livestock and Land Use Surveys for Narok, Kajiado and Kitui Districts." Prepared for the Government of Kenya, Nairobi, 1975.

Werbner, R. "Introduction." Land Reform in the Making: Tradition, Public Policy and Ideology in Botswana, ed. R. Werbner. Special Number of the Journal of African Law 24 (Spring 1980):i-xii.

Western, D. "The Environment and Ecology of Pastoralists in Arid Savannas." Development and Change 13 (April 1982): 183-211.

White, G. "Natural Hazards Research: Concepts, Methods, and Policy Implications." Natural Hazards, ed. G. White, pp. 3-16. New York: Oxford University Press, 1974.

White, J.M. "The Role of the Livestock Marketing Division in Cattle Marketing in Kenya." Livestock Development Projects Course, sponsored jointly by ILCA and the Economic Development Institute, IBRD, Nairobi, 1978.

White, J.M., and S.J. Meadows. "Meat Supplies." Ministry of Agriculture, Nairobi, 1978.

White, J.M., and S.J. Meadows. "The Potential Supply of Immatures over the 1980s from Kenya's Northern Rangelands." Ministry of Agriculture, Nairobi, 1980a.

White, J.M., and S.J. Meadows. "Cattle Herd Composition and Milk Recording Exercises, Southern Rangelands." Ministry of Agriculture, Nairobi, 1980b.

White, J.M., and S.J. Meadows. "Evaluation of the Contribution of Group and Individual Ranches in Kajiado District to Economic and Social Development." Ministry of Livestock Development, Nairobi, 1981.

Wilemski, E. African Traditional Subsistence Economy in Change. Cologne: R.J. Hundt, 1975.

Wilson, R.T. "The Economic and Social Importance of Goats and Their Products in the Semi-Arid Arc of Northern Tropical Africa." Proceedings, Third International Conference on Goat Production and Disease, pp. 186-95. Scottsdale, Arizona: Dairy Goat Journal, 1982.

Wilson, R.T., C. Peacock, and A.R. Sayers. "A Study of Goat and Sheep Production of the Maasai Group Ranch at Elangata Wuas, Kajiado District, Kenya." Addis Ababa, Ethiopia: ILCA Arid and Semi-Arid Zones Programme Document No. 56, 1981.

Winrock International. "Livestock Program Priorities and Strategy." Draft position paper prepared by Winrock International Livestock Research and Training Center for U.S.A.I.D., Contract 53-319R-1-202, Morrilton, Arkansas, 1981.

Wyeth, P. "Economic Development in Kenyan Agriculture." Papers on the Kenyan Economy: Performance, Problems and Policies, ed. T. Killick, pp. 299-310. Nairobi, Kenya: Heinemann Educational Books (East Africa) Ltd., 1981.

Zeleny, M. "Compromise Programming." Multiple Criteria Decision Making, ed. J. Cochrane and M. Zeleny, pp. 262-301. Columbia: University of South Carolina Press, 1973.

Index

For Product Safety Concerns and Information please contact our EU representative GPSR@taylorandfrancis.com Taylor & Francis Verlag GmbH, Kaufingerstraße 24, 80331 München, Germany

Batch number: 10725877

Printed by Printforce, the Netherlands